Lesson Plans

Teaching Companion
Guide to Green Fabrics

Published by:

Kristene Smith, Incorporated
Post Office Box 233553
Sacramento, California 95823

Library of Congress Control Number 2011909134

ISBN-13 9780989474719

First Edition

Guide to Green Fabrics Teaching Companion™

Table of Contents

Introduction

How to Use the Guide to Green Fabrics Teaching Companion™

Instructors are encouraged to utilize the Guide to Green Fabrics Teaching Companion™ as a means to quickly and efficiently deliver ready-made lesson plans to students. Each chapter in Guide to Green Fabrics™ relates to the thirty-two (32) pre-designed lesson plans, and each lesson also contains a chapter quiz to help students study and retain new information they learn about eco-fibers. To enhance learning and to help students follow the lesson plans, students should be encouraged to purchase the Guide to Green Fabrics Student Workbook™, whereby the chapter quizzes contained therein should be completed and reviewed weekly, or as lessons are assigned. Completion of the quizzes contained within the Student Workbook could be counted toward the student's overall grade (percentage of final grade), or as extra credit resulting in a number of points the student can earn to offset the overall grade.

This Teaching Companion is divided into six (6) sections. Below you will find detail on how to fully utilize each section:

Section 1: **Pre-Midterm Lesson Plans**

The pre-midterm lesson plans are designed to provide students with a rich, introductory base of knowledge surrounding the world of eco-fibers, and builds upon prior learning students may have been exposed to. You will notice that the lesson plans are structured with a distinctive flow of information and are similar in tone from chapter to chapter. The amount of detail contained in each chapter will assist students in understanding how green fibers contribute to the environment and help shape the way various industries continue to integrate sustainable materials into design.

Section 2: **Midterm Assignment Options**

Within the early lesson plans time is allotted for the midterm overview as is student working time for the actual midterm assignment. The midterm includes two options: *Films that Influence*, and a *Research Presentation Project*. For your convenience, reproducible midterm worksheets (Midterm Assignment Options – Options 1 and 2) are provided in Section 6 and both contain suggested grading criteria and can be used as cover sheets to the actual midterm work. We also suggest allowing ten (10) minute class presentations for students who choose to participate in the *Research Presentation Project*. Although suggested to take place within certain lesson plans, these presentations should take place during times that accommodate your teaching schedule.

Section 3: **Post-Midterm Lesson Plans**

Lesson plans contained in this section further student learning and exposure to green fabrics. This section explores organic and recycled fibers, among others, and gives students a greater understanding of how the textiles industry operates, and how sustainability continues to be integrated into traditional production methods.

Section 4: **Answer Key**

The answer key is located in Section 4 and provides answers to all three parts of the final exam (questions, term definitions, and multiple-choice questions). In addition, answers are provided for Extra Credit Assignments # 6 through # 19.

Section 5: **Suggested Grading Criteria**

For your reference and consideration, suggested final grade grading criteria is provided in Section 5.

Section 6: **Reproducible Worksheets (enclosed)**

Reproducible worksheets are provided for the convenience of the instructor, to save time and other resources in delivering ready-made assignments to students. The value of a separate package (Section 6) with single-sided sheets also aids teachers in quickly copying and disseminating worksheets to students. This section contains three separate sets of worksheets: Part I, Unit Test/ Final Exam; Part II, Midterm Assignment Options – Options 1 and 2; and Part III, Extra Credit Assignments. At your discretion, students may earn extra credit by completing the Extra Credit Assignments perhaps up to a maximum, thereby using those points toward their final grade. Please note that the Student Workbook contains chapter lessons and references a midterm assignment, but does not contain any worksheets for students. The book, Guide to Green Fabrics™, the Teaching Companion, the Student Workbook, and Power Points are designed to work together with the instructor having access to the master teaching tools that advance student learning. All assignments contained within are provided as suggested guidelines and can be modified to suit your teaching style and needs.

Lesson Plans

Teaching Companion
Guide to Green Fabrics

Section 1 | Pre-Midterm Lesson Plans

Abaca - Kenaf

Chapter 1 | **Abaca** *Lesson One*

Introduction – Three Things You Know

- Have students write down three things they know about abaca.
- Have them share their knowledge with the class and make a list on the board.
- Try to answer any questions they have and facilitate any discussion that happens.
- Tell the students about the assigned reading and that you are going to show them a Power Point presentation and that they must write down three things they learn from both of these exercises. Have them save their answers to share later.

Reading

- Allow reading time for the abaca chapter in class.
- Have students finish reading the abaca chapter at home.

Power Point Presentation

- Show the class the abaca chapter Power Point presentation.
- Talk through each slide and explain the contents.
- Students are to write down three things they learned.
- Answer any questions as best you can and foster discussion.

Mass Media

- Have students break into groups to discuss how green fabrics and other environmental concerns are portrayed in the mass media. Assign each group the task of bringing to class an article, print ad, online promotion, film review, or advertisement showing how green fabrics, green products, and related concerns are promoted or portrayed.

Quick Answer: Sustainability Approach

- During class, ask students the following question in an open discussion format: What would be your approach to developing an innovative solution to the following: consumer waste?

Three Things You've Learned

- Students will review their original abaca-related entries and compare them to what they've learned during the reading and Power Point exercises.

Abaca Chapter Quiz

- Students will take the abaca chapter quiz located within the Student Workbook.

Abaca Chapter Quiz

Q1: In what way does abaca differ from true hemp? List the various differences between the two varieties.

A1: True hemp is a soft fiber (a product of Cannabis sativa) whereas abaca is a hard fiber.

Q2: How has abaca impacted the economic fortunes of the Philippines?

A2: Having consistent, positive impact on the region, abaca has helped in the socio-economic advancement of the people of the Philippines. It is recognized as its flagship commodity, especially of the Eastern Visayas Region. Grown in 48 provinces, Cantanduanes is abaca's top producer. The nine others are Leyte, Southern Leyte, Northern Samar, Davao Oriental, Surigao del Sur, Davao del Sur, Sulu, Sorsogon and Western Samar. The abaca plant has been introduced into many different tropical regions including Indonesia, the Andaman Islands in India, the West Indies, and Central America. In Ecuador, abaca is grown on large estates, much in contrast to the small farm production common to the Philippines.

Q3: While the use of abaca in the fashion and furnishings industries is common, how does it aid people suffering from diseases like psoriasis and eczema?

A3: Some therapeutic skin care products that are hand-crafted and sold as specialty items now contain abaca due to its high enzyme content. It is being added to cosmetics, soaps, and beauty creams throughout the world. Europe, the United States, and Japan are producers and marketers of such products, which reportedly have age defying properties that attract consumers worldwide. In treating eczema and psoriasis, it is especially effective as it moisturizes and helps in skin regeneration.

Q4: Grading of abaca fiber depends on certain production and drying techniques. List these.

A4: For purposes of grading, the part of the stem from where it is obtained, the amount of serration on the stripping knives, and the degree of tension while holding the knife against the block are taken into account to determine the quality of the fiber. Promptness and care taken while drying also affect the grade and quality of the fiber. In Central America, a widely used mechanical method involves cutting the stalks and crushing and scraping them by machine, then drying the fibers mechanically.

Chapter 2 | **Bamboo** *Lesson Two*

Introduction – Three Things You Know

- Have students write down three things they know about bamboo.
- Have them share their knowledge with the class and make a list on the board.
- Try to answer any questions they have and facilitate any discussion that happens.
- Tell the students about the assigned reading and that you are going to show them a Power Point presentation and that they must write down three things they learn from both of these exercises. Have them save their answers to share later.

Reading

- Allow reading time for the bamboo chapter in class.
- Have students finish reading the bamboo chapter at home.

Power Point Presentation

- Show the class the bamboo chapter Power Point presentation.
- Talk through each slide and explain the contents.
- Students are to write down three things they learned.
- Answer any questions as best you can and foster discussion.

Open Discussion

- Discuss the following with the class in an open format: In terms of cultural exchange, education, and the passing of traditions, how can international governments work together to cross-train workers on the special skills needed to harvest and locally manufacture regionally grown sustainable fibers? Would this be viable? Does this pose problems relating to local economic development especially in developing nations? Remember to consider language barriers and *secrets of the trade* in your response.

Quick Answer: Sustainability Approach

- During class, ask students the following question in an open discussion format: What would be your approach to developing an innovative solution to the following: eco-messaging and advertising?

Three Things You've Learned

- Students will review their original bamboo-related entries and compare them to what they've learned during the reading and Power Point exercises.

Bamboo Chapter Quiz

- Students will take the bamboo chapter quiz located within the Student Workbook.

Bamboo Chapter Quiz

Q1: Bamboo cultivation in China is extremely eco-friendly. Why?

A1: In China, organic bamboo is cultivated on over 7 million acres of family-owned farms, therefore posing no threat to tropical rain forests in other locales.

Q2: Clothes made of bamboo are also referred to as *air conditioning dress*. What unique properties make the dress cool and suitable for summer?

A2: One of its signature attractions is its breathability created by the micro-gaps and micro-holes that are present in the fiber. These tiny vents allow for increased air circulation, and its wick-like properties enhance its absorption qualities. Even at the height of summer a fashionista attired in garments made of bamboo can remain cool and chic knowing well that the garment will hold its appeal. This feature has earned the fiber a special nickname: *air-conditioning dress*.

Q3: Bamboo viscose can mimic several other fibers and also does not use formaldehyde, which is a known carcinogen. Name the fibers it mimics and the quality of bamboo viscose which makes it an apt replacement for these fibers.

A3: With respect to its chameleon-like characteristics, bamboo viscose (rayon) is a leader in this category. It has the ability to mimic other fibers such as wool, silk, linen and rayon. It can be woven or knit into any type of fabric from denim to a slinky jersey. Other options include silky bamboo, bamboo canvas, double knit, terry cloth, shirting, and a variety of bamboo yarns for knitting, weaving, and crocheting. These fabrics generally resist wrinkling without the addition of formaldehyde (a carcinogen), which is added to polyester/ cotton blended fabrics to give them those qualities. However, all wrinkling is not eliminated, especially for wovens.

Q4: What quality of bamboo fiber makes it ideal for medical and health related products?

A4: Bamboo kun, which is the anti-bacterial component of bamboo, comes from the presence of a bio-agent, which is bound tightly to bamboo's cellulose molecular structure. This bacteriostasis agent arrests the growth and spread of odor causing bacteria giving bamboo fabric a feeling of freshness all day long. Even after 50 washes the fabric retains its anti-fungal, anti-bacterial nature, depending on the production process.

Chapter 3 | Cervelt™ *Lesson Three*

Introduction – Three Things You Know

- Have students write down three things they know about Cervelt™.
- Have them share their knowledge with the class and make a list on the board.
- Try to answer any questions they have and facilitate any discussion that happens.
- Tell the students about the assigned reading and that you are going to show them a Power Point presentation and that they must write down three things they learn from both of these exercises. Have them save their answers to share later.

Reading

- Allow reading time for the Cervelt™ chapter in class.
- Have students finish reading the Cervelt™ chapter at home.

Power Point Presentation

- Show the class the Cervelt™ chapter Power Point presentation.
- Talk through each slide and explain the contents.
- Students are to write down three things they learned.
- Answer any questions as best you can and foster discussion.

Celebrity Profile

- Using the internet and other resources, have students choose an eco-minded celebrity to profile in a short, yet complete, one-page report. Instruct students to include the reasons why the celebrity is eco-minded and have them list some of the celebrity's associations to the sustainability community.

Three Things You've Learned

- Students will review their original Cervelt™-related entries and compare them to what they've learned during the reading and Power Point exercises.

Cervelt™ Chapter Quiz

- Students will take the Cervelt™ chapter quiz located within the Student Workbook.

Cervelt™ Chapter Quiz

Q1: Why is Cervelt™ considered an extremely expensive and rare eco-friendly fiber?

A1: Cervelt™, defined as *the diamond of clothing*, is a new and spectacularly eco-driven luxury fiber. It is a rare fiber obtained from the red deer of New Zealand, which boasts the world's largest population of this prized creature. Only a small amount of fiber can be taken from one of the deer, about 20 grams maximum, therefore, it remains rare.

Q2: Cervelt™ remains pill free. What qualities of the fiber allow it to remain so?

A2: Cervelt™ fashions are pill-free due to the fiber's coil-like qualities, which add lightness, softness, and elasticity.

Q3: Cervelt™ is made from raw material obtained from the red deer. Where is this breed of deer found?

A3: This breed of red deer are from New Zealand.

Q4: What conditions are made available to the red deer to ensure its long life?

A4: Extraordinarily well cared for, these deer are provided with spacious, beautiful grounds to roam naturally and stress-free. They drink unpolluted river water, breathe pure air, and nestle amongst unmatched, tranquil scenery, thereby ensuring their long life.

Chapter 4 | Coir

Lesson Four

Introduction – Three Things You Know

- Have students write down three things they know about coir.
- Have them share their knowledge with the class and make a list on the board.
- Try to answer any questions they have and facilitate any discussion that happens.
- Tell the students about the assigned reading and that you are going to show them a Power Point presentation and that they must write down three things they learn from both of these exercises. Have them save their answers to share later.

Reading

- Allow reading time for the coir chapter in class.
- Have students finish reading the coir chapter at home.

Power Point Presentation

- Show the class the coir chapter Power Point presentation.
- Talk through each slide and explain the contents.
- Students are to write down three things they learned.
- Answer any questions as best you can and foster discussion.

Creative Worksheet

- Fashion Design. Have students sketch a green fashion design with the personality of a known celebrity. Students must submit a full color sketch or painting for this assignment.

Quick Answer: Sustainability Approach

- During class, ask students the following question in an open discussion format: What would be your approach to developing an innovative solution to the following: natural dye and coloring?

Three Things You've Learned

- Students will review their original coir-related entries and compare them to what they've learned during the reading and Power Point exercises.

Coir Chapter Quiz

- Students will take the coir chapter quiz located within the Student Workbook.

Coir Chapter Quiz

Q1: Elaborate on the connection between coir and India.

A1: One of the oldest fibers known to man, coir is a product of the coconut, one of the earth's first trees, and its intrinsic association with India arises not only because it is found in abundance there, but is also an important part of Indian mythology and rituals. In Indian mythology the tree is considered as a *kalpa-vriksha*, or the *wish-fulfilling tree*, and in modern interpretation it can be considered so because it fulfills the economic and material aspirations of those who cultivate it.

Q2: How tall can coconut trees grow?

A2: These trees can grow up to an amazing 100 feet tall.

Q3: How did colonialism help the spread of coir fiber and its usage?

A3: With the advent of colonialism coir was soon traveling to various parts of the globe because the various European countries that colonized India sent it to their mother country and other colonies to be used for ropes and mats.

Q4: Which well-known carpet firm used coir in fabric production?

A4: The very well-known carpet firm, Treloar and Sons, used coir in fabric production.

Chapter 5 | Corn Husk
Lesson Five

Introduction – Three Things You Know

- Have students write down three things they know about corn husk.
- Have them share their knowledge with the class and make a list on the board.
- Try to answer any questions they have and facilitate any discussion that happens.
- Tell the students about the assigned reading and that you are going to show them a Power Point presentation and that they must write down three things they learn from both of these exercises. Have them save their answers to share later.

Reading

- Allow reading time for the corn husk chapter in class.
- Have students finish reading the corn husk chapter at home.

Power Point Presentation

- Show the class the corn husk chapter Power Point presentation.
- Talk through each slide and explain the contents.
- Students are to write down three things they learned.
- Answer any questions as best you can and foster discussion.

Open Discussion

- Discuss the following with the class in an open format: How can the design industry improve the appeal of eco-products in general? With the sheer cost of green design being a factor, what messages and techniques can be employed using mass communications and consumer advertising?

Three Things You've Learned

- Students will review their original corn husk-related entries and compare them to what they've learned during the reading and Power Point exercises.

Corn Husk Chapter Quiz

- Students will take the corn husk chapter quiz located within the Student Workbook.

Corn Husk Chapter Quiz

Q1: What are the similarities between corn husk fiber, cotton, and linen?

A1: Corn husk fiber exhibits properties that mimic those of cotton and linen in strength, while the length of the fiber is more like that of cotton. It is softer and more flexible than both cotton and linen with greater load bearing capacity, imparting strength to the fabric.

Q2: Name one of the most important products made using corn husk.

A2: One of the most important products made using corn husks is twine and this was prompted by the poor economics of growing and producing sisal fiber and twine in the U.S., which was the preferred material for binding bales of straw. The climatic conditions and immense labor requirements rendered sisal cultivation non-viable in the U.S. leading to the discovery of creating fiber from husks.

Q3: Why is corn husk fiber considered an extremely green option in spite of the chemicals used in producing the fiber?

A3: The impact on the environment has been positive considering that the utilization of waste product to create a material of great utility is considered noteworthy within the environmental community. Production of the fiber is extremely simple, and without excess mechanical input, this makes it not only a green option, but also an economically viable one.

Q4: Why is corn husk considered greener than traditional cotton?

A4: Corn cultivation calls for far fewer fertilizers and pesticides when compared to traditional textile fibers like cotton.

Chapter 6 | **Eco-Fi®** *Lesson Six*

Introduction – Three Things You Know

- Have students write down three things they know about EcoFi®.
- Have them share their knowledge with the class and make a list on the board.
- Try to answer any questions they have and facilitate any discussion that happens.
- Tell the students about the assigned reading and that you are going to show them a Power Point presentation and that they must write down three things they learn from both of these exercises. Have them save their answers to share later.

Reading

- Allow reading time for the EcoFi® chapter in class.
- Have students finish reading the EcoFi® chapter at home.

Power Point Presentation

- Show the class the EcoFi® chapter Power Point presentation.
- Talk through each slide and explain the contents.
- Students are to write down three things they learned.
- Answer any questions as best you can and foster discussion.

Social/ Mobile

- During class, ask students the following question in an open discussion format: Explain how social media and mobile technology are empowering green designers to spread messages of environmental awareness across the globe. Elaborate its contribution in building a global community that supports green fabrics and makes efforts to promote it among peers and colleagues. Are these results better as compared to conventional media such as print and television? How so?

Quick Answer: Sustainability Approach

- During class, ask students the following question in an open discussion format: What would be your approach to developing an innovative solution to the following: marketing green fabrics?

Three Things You've Learned

- Students will review their original EcoFi®-related entries and compare them to what they've learned during the reading and Power Point exercises.

Eco-Fi® Chapter Quiz

- Students will take the EcoFi® chapter quiz located within the Student Workbook.

EcoFi® Chapter Quiz

Q1: Compared to natural fibers, what are two of the advantages of EcoFi®?

A1: EcoFi® has some advantages not found in natural fibers, the most important being extreme colorfastness and lack of shrinkage.

Q2: Enumerate some of the benefits of using EcoFi® relative to its care.

A2: EcoFi® is relatively easy to take care of since the manufacturing process itself manages attendant difficulties found with natural fibers. It is machine washable and dryable. Garments created from this fabric can be laundered by following manufacturers' washing instructions.

Q3: Which company won the first UN Fashion Industry Award for Environmental Excellence?

A3: Wellman, Inc., the manufacturer of EcoFi®, so impressed the United Nations, one of the foremost crusaders of environmental protection, that it was awarded the first Fashion Industry Award for Environmental Excellence.

Q4: What qualities of EcoFi® make it a fabric perfect for use in outdoor sports and other allied activities?

A4: EcoFi® fabric features very good insulation properties, and for this reason, it is most commonly used in creating sports clothing, especially those for outdoor activities. Its capacity to retain heat makes it perfect for the manufacture of thermal undergarments, sweaters, and socks.

Chapter 7 | **Eco-Leather/ E-Leather** *Lesson Seven*

Introduction – Three Things You Know

- Have students write down three things they know about eco-leather/ e-leather.
- Have them share their knowledge with the class and make a list on the board.
- Try to answer any questions they have and facilitate any discussion that happens.
- Tell the students about the assigned reading and that you are going to show them a Power Point presentation and that they must write down three things they learn from both of these exercises. Have them save their answers to share later.

Reading

- Allow reading time for the eco-leather/ e-leather chapter in class.
- Have students finish reading the eco-leather/ e-leather chapter at home.

Power Point Presentation

- Show the class the eco-leather/ e-leather chapter Power Point presentation.
- Talk through each slide and explain the contents.
- Students are to write down three things they learned.
- Answer any questions as best you can and foster discussion.

My Own Way

- Ask students if they have ever been in a situation whereby they were confronted with challenges to their eco-friendly beliefs. What did they do and how did they and others react? This assignment can be discussed in an open format with the entire class.

Quick Answer: Sustainability Approach

- During class, ask students the following question in an open discussion format: What would be your approach to developing an innovative solution to the following: organic farming methods?

Three Things You've Learned

- Students will review their original eco-leather/ e-leather-related entries and compare them to what they've learned during the reading and Power Point exercises.

Eco-Leather/ E-Leather Chapter Quiz

- Students will take the eco-leather/ e-leather chapter quiz located within the Student Workbook.

Eco-Leather/ E-Leather Chapter Quiz

Q1: What is the difference between conventionally processed leather and eco-leather?

A1: Conventionally processed leather uses numerous chemicals in production and processing, which causes damage to the environment. Today, conventional tanning methods using chrome are being replaced with healthy and environmentally-friendly vegetable tanning methods. Eco-leather, which as the name suggests, is friendly to the environment as it uses natural processing methods, leaving the curing and coloring to nature and natural dyes.

Q2: What makes South America one of the largest providers of raw material for eco-leather?

A2: Hides used to make eco-leather come mostly from South America, as the cows that are reared in this region are organic in nature. These cows are not certified as organic, but since they graze freely on huge expanses of pasture that grow naturally, their organic label is pre-determined.

Q3: What is a *growth mark* and how does it help in establishing the identity of eco-leather?

A3: A *growth mark* is visible in a healed scar, which is the result of damage caused by barbed wire or horns to cattle from whom leather is used for creating eco-leather products. The age of the animal from which the hide is taken can also be deciphered by growth marks and veins.

Q4: What happens to growth marks over time?

A4: With the passage of time growth marks develop a patina that adds to its natural beauty giving it distinct characteristics.

Chapter 8 | **Hemp** *Lesson Eight*

Introduction – Three Things You Know

- Have students write down three things they know about hemp.
- Have them share their knowledge with the class and make a list on the board.
- Try to answer any questions they have and facilitate any discussion that happens.
- Tell the students about the assigned reading and that you are going to show them a Power Point presentation and that they must write down three things they learn from both of these exercises. Have them save their answers to share later.

Reading

- Allow reading time for the hemp chapter in class.
- Have students finish reading the hemp chapter at home.

Power Point Presentation

- Show the class the hemp chapter Power Point presentation.
- Talk through each slide and explain the contents.
- Students are to write down three things they learned.
- Answer any questions as best you can and foster discussion.

News Article Review

- Have students research a current news article on eco-friendly fabrics. Ask students to describe its contents in detail and provide insights to the class about its relevancy to sustainability.

Quick Answer: Sustainability Approach

- During class, ask students the following question in an open discussion format: What would be your approach to developing an innovative solution to the following: corporate subsidies for eco-farmers?

Three Things You've Learned

- Students will review their original hemp-related entries and compare them to what they've learned during the reading and Power Point exercises.

Hemp Chapter Quiz

- Students will take the hemp chapter quiz located within the Student Workbook.

Hemp Chapter Quiz

Q1: Hemp has a negative connotation because of its association with which plant?

A1: Marijuana.

Q2: In what way is hemp associated with the Declaration of Independence and the Founding Fathers of the United States?

A2: The Declaration of Independence was first drafted on paper made from hemp while Benjamin Franklin's paper mill made paper only from hemp. George Washington also cultivated hemp.

Q3: Elaborate on the connection between the U.S. military and hemp.

A3: The strength, durability, and low cost of hemp made it the fabric of choice for army uniforms, and Napoleon's soldiers were outfitted in hemp uniforms. During World War II, the U.S. Army used hemp for uniforms, rope, and canvas. It was used in ships to secure sails, as it is highly resistant to saltwater rot and thus ideal for sea-going vessels.

Q4: Describe the climatic and soil conditions ideal for cultivating hemp.

A4: Hemp can grow in a range of soils, but tends to prefer those that produce high yields of corn. Ideally, the soil would be well drained, non-acidic, and rich in nitrogen. Mild, humid climates with an average rainfall of at least 25 to 30 inches (64-76 cm) per year suit it best.

Q5: Hemp is a bast fiber. What is a bast fiber?

A5: Bast fibers are fibers which are extracted from the soft part of a plant found just outside the woody inners of the stalk.

Q6: What is *retting* and what is the difference between traditional and modern retting methods?

A6: *Retting* is the process of soaking a cut plant in water for several days to loosen the fibers, after which the stems are bundled up into small rolls and passed through a series of rollers to get rid of any leftover wood and stem. Alternately, modern retting processes can vary from using enzymes and chemicals in order to hasten the process, to using ultrasound technology.

Chapter 9 | Henequen/ Sisal

Lesson Nine

Introduction – Three Things You Know

- Have students write down three things they know about henequen/ sisal.
- Have them share their knowledge with the class and make a list on the board.
- Try to answer any questions they have and facilitate any discussion that happens.
- Tell the students about the assigned reading and that you are going to show them a Power Point presentation and that they must write down three things they learn from both of these exercises. Have them save their answers to share later.

Reading

- Allow reading time for the henequen/ sisal chapter in class.
- Have students finish reading the henequen/ sisal chapter at home.

Power Point Presentation

- Show the class the henequen/ sisal chapter Power Point presentation.
- Talk through each slide and explain the contents.
- Students are to write down three things they learned.
- Answer any questions as best you can and foster discussion.

Essay

- Have students write an 800-word essay on the impact of green design on the media. How prevalent is green design, and are there enough media worthy topics emanating from the industry to generate coverage? Which topics generate the most interest? Remind students to use cited sources in their responses.

Quick Answer: Sustainability Approach

- During class, ask students the following question in an open discussion format: What would be your approach to developing an innovative solution to the following: organic crop pest control?

Three Things You've Learned

- Students will review their original henequen/ sisal-related entries and compare them to what they've learned during the reading and Power Point exercises.

Henequen/ Sisal Chapter Quiz

- Students will take the henequen/ sisal chapter quiz located within the Student Workbook.

Henequen/ Sisal Chapter Quiz

Q1: How is the economy of Mexico tied with sisal or henequen production?

A1: Henequen has almost always been synonymous with the Yucatan region of Mexico and the economy of Yucatan owes much to this strong, sturdy, resilient fiber which was among its biggest and most important cash crops. The wealth derived from henequen created some of the Yucatan's most magnificent haciendas and made this Mexican state one of the most prosperous in the country.

Q2: Elaborate on the link between the invention of the reaper and the use of sisal ropes.

A2: The biggest impetus to the henequen industry came when the reaper was invented by American Cyrus McCormick. This machine was used to cut standing grain that was then tied into bales using thin wire, which was often ingested by cows piercing their stomachs and making them sick. When nylon ropes were used to tie the bales they too were ingested by the cows leading to sickness. It was then that henequen ropes were used to tie bales of cattle feed, and being a natural fiber, even upon ingestion by the animals, was easily digested and did not cause harm.

Q3: What feature of the sisal plant makes it a favored crop in developing and third world nations?

A3: Some of the countries engaged in producing the fiber in significant quantities are Brazil, Tanzania, Philippines, Madagascar, Venezuela, and Florida in the United States. Being a very tough plant, it survives in even the most hostile of environments and while several other crops cannot survive a drought, henequen remains unaffected by lack of rainfall. This makes it especially popular as a cash crop in developing countries where irrigation systems are relatively less developed and cultivators have to depend on rainfall for water.

Q4: The use of a certain chemical makes sisal fiber fire retardant. Name the chemical.

A4: Henequen fabric can also serve as a fire retardant after it has been treated with borax.

Chapter 10 | Ingeo
Lesson Ten

Introduction – Three Things You Know

- Have students write down three things they know about ingeo.
- Have them share their knowledge with the class and make a list on the board.
- Try to answer any questions they have and facilitate any discussion that happens.
- Tell the students about the assigned reading and that you are going to show them a Power Point presentation and that they must write down three things they learn from both of these exercises. Have them save their answers to share later.

Reading

- Allow reading time for the ingeo chapter in class.
- Have students finish reading the ingeo chapter at home.

Power Point Presentation

- Show the class the ingeo chapter Power Point presentation.
- Talk through each slide and explain the contents.
- Students are to write down three things they learned.
- Answer any questions as best you can and foster discussion.

Celebrity Profile

- Using the internet and other resources, have students choose an eco-minded celebrity to profile in a short, yet complete, one-page report. Instruct students to include the reasons why the celebrity is eco-minded and have them list some of the celebrity's associations to the sustainability community.

Three Things You've Learned

- Students will review their original ingeo-related entries and compare them to what they've learned during the reading and Power Point exercises.

Ingeo Chapter Quiz

- Students will take the ingeo chapter quiz located within the Student Workbook.

Ingeo Chapter Quiz

Q1: What does the term *Ingeo* mean?

A1: Ingeo means *ingredients from the earth* and it is one of the finest man made fabrics not sourced from synthetic materials.

Q2: Name some of the qualities that ingeo shares with synthetic fibers.

A2: Like all synthetics, the fibers come equipped with good strength, high tensile properties, and good resistance to wear and tear. Ingeo has the added advantage of being less susceptible to ultraviolet rays, one of the main causes of radiation brought about by ozone layer depletion.

Q3: Which countries are the main producers of ingeo?

A3: South Korea, Taiwan, and Mainland China are the leading producers of ingeo.

Q4: Describe the method used for producing ingeo fiber. What is ingeo used for?

A4: Ingeo is the result of chemical processing involving procedures of basic fermentation, distillation, and subsequent polymerization. Naturally occurring starches found in plants such as corn are broken down into sugar, which is made into a biopolymer (PLA). The resin is then used to create plastic goods via injection molding and thermoformed into packaging. It is also extruded for film and textile applications.

Chapter 11 | Jute
Lesson Eleven

Introduction – Three Things You Know

- Have students write down three things they know about jute.
- Have them share their knowledge with the class and make a list on the board.
- Try to answer any questions they have and facilitate any discussion that happens.
- Tell the students about the assigned reading and that you are going to show them a Power Point presentation and that they must write down three things they learn from both of these exercises. Have them save their answers to share later.

Reading

- Allow reading time for the jute chapter in class.
- Have students finish reading the jute chapter at home.

Power Point Presentation

- Show the class the jute chapter Power Point presentation.
- Talk through each slide and explain the contents.
- Students are to write down three things they learned.
- Answer any questions as best you can and foster discussion.

Mass Media

- Have students break into groups to discuss how green fabrics and other environmental concerns are portrayed in the mass media. Assign each group the task of bringing to class an article, print ad, online promotion, film review, or advertisement showing how green fabrics, green products, and related concerns are promoted or portrayed.

Quick Answer: Sustainability Approach

- During class, ask students the following question in an open discussion format: What would be your approach to developing an innovative solution to the following: green product packaging design?

Three Things You've Learned

- Students will review their original jute-related entries and compare them to what they've learned during the reading and Power Point exercises.

Jute Chapter Quiz

- Students will take the jute chapter quiz located within the Student Workbook.

Jute Chapter Quiz

Q1: Apart from its color, why is jute called the *golden fiber*? Remember to consider jute's many levels of uses and appeal in your answer.

A1: From seed to the expired fiber, there is no part of the plant that can be cast aside as waste material. Jute gives only main products and extremely useful by-products, none of which are wasted. The uses of jute range from the common rope to designer jute silk that graces the homes of the rich and famous, and of course, the green brigade.

Its subtle earthiness makes it the perfect complement to designer homes that are now created around the back-to-nature theme. From drapes to upholstery, from rugs to throws and floor mats to place mats, jute fulfills all the accessorizing needs of modern homes in keeping with best environmental practices, earning it the name *golden fiber*.

Q2: Which natural substance found in jute aids in the attachment of short jute fibers to form longer ones?

A2: Elementary jute fibers are short in length measuring between 0.7 to 6 mm while the longer fibers, 3 to 4 cm long, are the result of the short fibers attaching themselves to each other through lignin.

Q3: Why is processing jute without chemicals a best practice, thus making it extremely environmentally-friendly?

A3: Jute is a complex mix of cellulose, hemicelluloses, and lignin. The hemicelluloses make it sensitive to alkali while the cellulose makes it sensitive to acid, rendering chemical processing of the fiber extremely delicate. Jute is also one of those rare cash crops that are the delight of environmentalists and economists alike. Being a crop grown in mostly developing countries, the methods of farming are still traditional, labor-intensive, and old-fashioned, which eliminates high use of chemicals and other pesticides.

Q4: During the retting process what action takes place in the plant's structure?

A4: During the retting process the phloem, which contains the useful fiber, is separated from the xylem, which is the woody core.

Chapter 12 | **Kenaf** *Lesson Twelve*

Introduction – Three Things You Know

- Have students write down three things they know about kenaf.
- Have them share their knowledge with the class and make a list on the board.
- Try to answer any questions they have and facilitate any discussion that happens.
- Tell the students about the assigned reading and that you are going to show them a Power Point presentation and that they must write down three things they learn from both of these exercises. Have them save their answers to share later.

Reading

- Allow reading time for the kenaf chapter in class.
- Have students finish reading the kenaf chapter at home.

Power Point Presentation

- Show the class the kenaf chapter Power Point presentation.
- Talk through each slide and explain the contents.
- Students are to write down three things they learned.
- Answer any questions as best you can and foster discussion.

Midterm Assignment Overview/ Sign-Up

Three Things You've Learned

- Students will review their original kenaf-related entries and compare them to what they've learned during the reading and Power Point exercises.

Kenaf Chapter Quiz

- Students will take the kenaf chapter quiz located within the Student Workbook.

Kenaf Chapter Quiz

Q1: What other names are used for kenaf?

A1: Kenaf, which is similar to hemp, is also called Deccan hemp and Java jute, and is extracted specifically from the plant Hibiscus Cannabinus, a plant in the Malvaceae family that is related to cotton.

Q2: What is kenaf's native country and in what other countries is it now produced?

A2: Kenaf is native to Africa, but gradually its cultivation has spread to many parts of the globe that have temperate to moist, tropical climates. It is a warm season row crop that grows annually. India is one of the largest producers while other countries where it is cultivated are Bangladesh, Malaysia, South Africa, Mexico, Italy, Spain, the United States, Russia, and the East Asian nations of China, Taiwan, and Indonesia.

Q3: What are the two methods commonly used to extract fiber from the kenaf plant? Provide detail in your response.

A3: Two methods of fiber extraction are followed: single plant extraction and the beat and jerk method. In the single plant extraction method fibers are separated from individual stems generally by hand and kept aside. The fibers obtained in this way are of a very superior quality. In the beat and jerk method about 10 to 12 plants are bundled together and the bottom portion pounded so as to loosen the fibers. The bundle is broken in the center, and with a final jerk, the fiber is removed from the stems.

Q4: Which country has labeled newsprint made from kenaf as *tree-free*?

A4: Accordingly, in 1960 the United States Department of Agriculture took into account the properties of over 500 plants and found kenaf to be the most promising source for newsprint and labeled it *tree-free*.

Lesson Plans

Teaching Companion
Guide to Green Fabrics

Section 2 | Midterm Assignment Options

Overview – Options 1 and 2

Overview | **Midterm Assignment** *Guide to Green Fabrics*

Overview:

Midterm assignments are designed to engage and enlighten students on global issues relating to the environment. Some of their research may not pertain directly to green fabrics, however, it is this level of global thinking that will propel students into greater awareness of how industries connect and how every individual can contribute to the success of positive environmental awareness efforts.

The assignments are broken into two categories: *Films that Influence* and a *Research Presentation Project*. Each will require significant study and organizing of ideas. You may also wish to develop other assignments to suit the learning styles of your students and your own teaching style. Read over the assignment options, discuss with the class, and based upon student preferences, distribute the midterm assignment worksheets to the class. Organize students into groups of three or four. You can assemble the groups randomly or by design, or you can have students select their own groups, or this can be an individual assignment for students not wanting to participate in a group. The grading criteria will relate to groups as well as individual students.

Midterm Assignment Options:

Midterm Assignment Option # 1 – Films that Influence (see Section 6 for worksheet)
Midterm Assignment Option # 2 – Research Presentation Project (see Section 6 for worksheet)

Midterm Assignment Overview/ Sign-up:

Give the groups a few minutes to confer then invite them to sign up for a topic – only one topic per group or individual.

Begin Research:

Students are to begin researching and taking notes about their topic. Make sure they understand how to take notes and that their presentations and visual aids must be original. Students may need access to the library and/or computers for this component.

Research Project Class Presentation:

Students will present their research presentation projects in ten (10) minute slots during future class periods.

Section 3

Post-Midterm Lesson Plans
Milk Protein – Wild Nettle

Chapter 13 | Milk Protein
Lesson Thirteen

Introduction – Three Things You Know
- Have students write down three things they know about milk protein.
- Have them share their knowledge with the class and make a list on the board.
- Try to answer any questions they have and facilitate any discussion that happens.
- Tell the students about the assigned reading and that you are going to show them a Power Point presentation and that they must write down three things they learn from both of these exercises. Have them save their answers to share later.

Reading
- Allow reading time for the milk protein chapter in class.
- Have students finish reading the milk protein chapter at home.

Power Point Presentation
- Show the class the milk protein chapter Power Point presentation.
- Talk through each slide and explain the contents.
- Students are to write down three things they learned.
- Answer any questions as best you can and foster discussion.

Midterm Worktime
- Students are to use this time to work on their midterm projects. Remind them to keep reading and researching at home.

Quick Answer: Sustainability Approach
- During class, ask students the following question in an open discussion format: What would be your approach to developing an innovative solution to the following: student eco-education?

Three Things You've Learned
- Students will review their original milk protein-related entries and compare them to what they've learned during the reading and Power Point exercises.

Milk Protein Chapter Quiz
- Students will take the milk protein chapter quiz located within the Student Workbook.

Milk Protein Chapter Quiz

Q1: Name the ancient beauty associated with the use of milk as a beauty aid. Name one of her favorite uses for milk.

A1: The skin improving qualities of milk have been known to man from ancient times and the beauty of Cleopatra was credited to the milk bath, one of her favorite indulgences. This may have been the secret behind her flawless complexion.

Q2: During which decade was milk protein fabric invented, and where? Give some of the registered names for fabrics made from milk protein.

A2: Milk protein fabric was invented in the 1930s in America and Italy, and was known as ARALAC®, Lanatil®, and Merinova®.

Q3: What fibers are commonly blended with milk protein fibers?

A3: Milk fibers are best used when blended with natural fibers such as cashmere, cotton, silk, wool, or ramie.

Q4: Explain *dewatering.*

A4: The process of turning liquid milk into fabric involves *dewatering* it, wherein the water content of the milk is removed to get it ready for skimming.

Chapter 14 | Modal
Lesson Fourteen

Introduction – Three Things You Know

- Have students write down three things they know about modal.
- Have them share their knowledge with the class and make a list on the board.
- Try to answer any questions they have and facilitate any discussion that happens.
- Tell the students about the assigned reading and that you are going to show them a Power Point presentation and that they must write down three things they learn from both of these exercises. Have them save their answers to share later.

Reading

- Allow reading time for the modal chapter in class.
- Have students finish reading the modal chapter at home.

Power Point Presentation

- Show the class the modal chapter Power Point presentation.
- Talk through each slide and explain the contents.
- Students are to write down three things they learned.
- Answer any questions as best you can and foster discussion.

Midterm Worktime

- Students are to use this time to work on their midterm assignments. This week begins the last time available to work on projects in class. Essays will simply be turned in, but for those choosing research presentations, they will begin presenting in class next week in structured or random order.

Quick Answer: Sustainability Approach

- During class, ask students the following question in an open discussion format: What would be your approach to developing an innovative solution to the following: recycling efforts?

Three Things You've Learned

- Students will review their original modal-related entries and compare them to what they've learned during the reading and Power Point exercises.

Modal Chapter Quiz

- Students will take the modal chapter quiz located within the Student Workbook.

Modal Chapter Quiz

Q1: What is the difference between rayon and modal fabric?

A1: A fabric similar to rayon, Modal is one of the newest fabrics to make a striking impact on the fashion and home décor industries. However, it differs from rayon in that rayon is produced from the chips of any tree, while Modal uses only Beech wood chips. It is a 100% cellulosic bio-fiber made completely by human ingenuity propelled by the desire to make maximum use of available resources found all around us.

Q2: What is the formal, registered name for modal?

A2: Modal is the industry name for modacrylic fiber. Its formal name is LenzingModal®, a registered trademark of Lenzing AG™, an Austrian company specializing in textiles and fibers, particularly natural fibers made from cellulose.

Q3: What negative aspect of natural fibers is eliminated in modal fabric?

A3: Modal is abrasion free and does not pill unlike natural fibers.

Q4: What does *hygroscopic* mean, and how is modal hygroscopic?

A4: Modal is highly hygroscopic, which means water absorbent. It does not lose its shape even upon becoming wet and will not shrink.

Chapter 15 | Natural Latex
Lesson Fifteen

Introduction – Three Things You Know
- Have students write down three things they know about natural latex.
- Have them share their knowledge with the class and make a list on the board.
- Try to answer any questions they have and facilitate any discussion that happens.
- Tell the students about the assigned reading and that you are going to show them a Power Point presentation and that they must write down three things they learn from both of these exercises. Have them save their answers to share later.

Reading
- Allow reading time for the natural latex chapter in class.
- Have students finish reading the natural latex chapter at home.

Power Point Presentation
- Show the class the natural latex chapter Power Point presentation.
- Talk through each slide and explain the contents.
- Students are to write down three things they learned.
- Answer any questions as best you can and foster discussion.

Midterm Presentations
- Students are to present their research projects to the class. Presentations are ten (10) minutes each and must be clearly presented both orally and visually. Remember to provide constructive feedback and individual and group evaluations.

Three Things You've Learned
- Students will review their original natural latex-related entries and compare them to what they've learned during the reading and Power Point exercises.

Natural Latex Chapter Quiz
- Students will take the natural latex chapter quiz located within the Student Workbook.

Natural Latex Chapter Quiz

Q1: What elements are found in latex which gives it its unique stickiness?

A1: Latex is the milky, sap-like fluid found in many plants that is a combination of several elements such as alkaloids, proteins, sugars, starch, resins, tannins, and gum.

Q2: Describe the journey of the rubber tree from Brazil to Southeast Asia.

A2: Brazil was a major producer and the originator of the rubber tree. In 1876, a local planter, Henry Wickham, exported seeds from Brazil to the United Kingdom on behalf of the British government. The seeds were germinated in London shortly thereafter. Eventually, seeds were exported to Sri Lanka, and in 1877, twenty-two seedlings made their way to Singapore, took root, and the tapping technique for latex extraction became standard. Natural latex harvesting and production mainly takes place in Thailand, Indonesia, Malaysia, and Sri Lanka, all located in Southeast Asia.

Q3: What is *bud grafting*?

A3: *Bud grafting* is a process by which buds from one plant are grafted onto another, and this is done so as to produce genetically identical trees in order to produce a pure product each and every time.

Q4: What is the *cambium* and why is it important to ensure that the cambium remains undamaged during the tapping process?

A4: In order for the tree to continue to grow in a healthy manner, the tree's cambium must remain intact. If it becomes damaged, the tree is considered damaged and will not grow nor produce latex since the cambium is where tree growth takes place. Hence, is it important to ensure that the cambium remains undamaged during the tapping process.

Chapter 16 | Organic Cashmere

Lesson Sixteen

Introduction – Three Things You Know

- Have students write down three things they know about organic cashmere.
- Have them share their knowledge with the class and make a list on the board.
- Try to answer any questions they have and facilitate any discussion that happens.
- Tell the students about the assigned reading and that you are going to show them a Power Point presentation and that they must write down three things they learn from both of these exercises. Have them save their answers to share later.

Reading

- Allow reading time for the organic cashmere chapter in class.
- Have students finish reading the organic cashmere chapter at home.

Power Point Presentation

- Show the class the organic cashmere chapter Power Point presentation.
- Talk through each slide and explain the contents.
- Students are to write down three things they learned.
- Answer any questions as best you can and foster discussion.

Midterm Presentations

- Students are to present their research projects to the class. Presentations are ten (10) minutes each and must be clearly presented both orally and visually. Remember to provide constructive feedback and individual and group evaluations.

Quick Answer: Sustainability Approach

- During class, ask students the following question in an open discussion format: What would be your approach to developing an innovative solution to the following: waterways and water supplies for organic farming?

Three Things You've Learned

- Students will review their original organic cashmere-related entries and compare them to what they've learned during the reading and Power Point exercises.

Organic Cashmere Chapter Quiz

- Students will take the organic cashmere chapter quiz located within the Student Workbook.

Organic Cashmere Chapter Quiz

Q1: From where did the word *cashmere* originate?

A1: Cashmere is an Anglicization of the word Kashmir, an Indian state, which is an important source of the wool.

Q2: How and when did cashmere find its way to Europe?

A2: Europeans were treated to the beauty and softness of this warm fabric when a French General, fighting in the French campaigns in Egypt during the wars between 1799 and 1812, sent home a shawl made of cashmere. From then on, the clamor for cashmere began, which continues unabated till date.

Q3: To whom is systematic cashmere production credited?

A3: Systematic production of the fabric is credited to Zyn-ul-Abidin, a 15th century Kashmir royal, who brought in weavers from Turkistan to create exotic designs.

Q4: How does the cashmere goat's *moulting* lead to the production of cashmere fiber? What is *dehairing*?

A4: When cashmere goats moult, shepherds get down to the arduous task of collecting the fiber and separating it from the coarse upper hair in a process called *dehairing*. This leads to the process of creating cashmere from the goat's hair.

Chapter 17 | Organic Cotton
Lesson Seventeen

Introduction – Three Things You Know

- Have students write down three things they know about organic cotton.
- Have them share their knowledge with the class and make a list on the board.
- Try to answer any questions they have and facilitate any discussion that happens.
- Tell the students about the assigned reading and that you are going to show them a Power Point presentation and that they must write down three things they learn from both of these exercises. Have them save their answers to share later.

Reading

- Allow reading time for the organic cotton chapter in class.
- Have students finish reading the organic cotton chapter at home.

Power Point Presentation

- Show the class the organic cotton chapter Power Point presentation.
- Talk through each slide and explain the contents.
- Students are to write down three things they learned.
- Answer any questions as best you can and foster discussion.

Creative Worksheet

- Interior Design. Have students develop a perspective (drawing) of a green interior design (single room only, any room choice) with the personality of a known consumer brand. Students must submit a full color drawing or painting for this assignment.

Quick Answer: Sustainability Approach

- During class, ask students the following question in an open discussion format: What would be your approach to developing an innovative solution to the following: consumer eco-education?

Three Things You've Learned

- Students will review their original organic cotton-related entries and compare them to what they've learned during the reading and Power Point exercises.

Organic Cotton Chapter Quiz

- Students will take the organic cotton chapter quiz located within the Student Workbook.

Organic Cotton Chapter Quiz

Q1: What is the difference between traditional and organic cotton?

A1: With traditional cotton production, there are issues of soil toxicity and run-off of harmful chemicals into waterways, among other environmental concerns. In fact, organic cotton production began as a result of these concerns. The harvesting and production of organic cotton is an extremely ethical enterprise, and the regulations that govern certification of cotton as organic are so stringent that total environmental benefits can be visibly maintained.

These innovative production methods ensure not only elimination of hazardous chemical substances, but at the same time replenish the soil making it more fertile and productive.

Q2: Which natural element of organic cotton is retained during processing lending to its elegant drape and fall?

A2: The fabric retains its natural wax resulting in its smooth hand, which in traditional cotton fabrics is stripped away during processing. This untreated fabric has more weight leading to elegant drape and excellent fall.

Q3: What precautions are taken in order to ensure the environmentally positive attributes of organic cotton at the planting and harvesting stages?

A3: For three years, organic fields must be cleansed and free of any prohibited substances prior to planting. Physical barriers and buzzers also need to be installed for the protection of organic crops, which must be maintained separately from chemically intensive crops and surface run-off. Soil fertility must also be promoted by farmers, and the biological, chemical, and physical condition of the soil should minimize erosion.

Q4: What factors led farmers to discontinue the practice of crop rotation?

A4: Many moons ago, farmers instinctively used crop rotation as a means of soil enrichment, fertility, retention, and pest control. In essence, pests, weeds, and disease could no longer anticipate the same plantings over and over, therefore, the soil became used to the receipt of organic matter and fresh crops that enriched it, freeing the soil from mundane and routine usage. As society advanced and economics became more of a factor in business and everyday life, farmers moved to a mono-crop culture meaning they planted single crops on vast amounts of land to yield more revenue. This discouraged crop rotation.

Chapter 18 | Organic Linen

Lesson Eighteen

Introduction – Three Things You Know

- Have students write down three things they know about organic linen.
- Have them share their knowledge with the class and make a list on the board.
- Try to answer any questions they have and facilitate any discussion that happens.
- Tell the students about the assigned reading and that you are going to show them a Power Point presentation and that they must write down three things they learn from both of these exercises. Have them save their answers to share later.

Reading

- Allow reading time for the organic linen chapter in class.
- Have students finish reading the organic linen chapter at home.

Power Point Presentation

- Show the class the organic linen chapter Power Point presentation.
- Talk through each slide and explain the contents.
- Students are to write down three things they learned.
- Answer any questions as best you can and foster discussion.

Open Discussion

- Discuss the following with the class in an open format: In order to propel early, eco-friendly education, should public and other primary education systems develop mandated, environmentally-focused, educational outreach programs for students? What benefits would this provide? What kinds of opposition to this proposal might occur, and why?

Quick Answer: Sustainability Approach

- During class, ask students the following question in an open discussion format: What would be your approach to developing an innovative solution to the following: government involvement and support of eco-fiber production?

Three Things You've Learned

- Students will review their original organic linen-related entries and compare them to what they've learned during the reading and Power Point exercises.

Organic Linen Chapter Quiz

- Students will take the organic linen chapter quiz located within the Student Workbook.

Organic Linen Chapter Quiz

Q1: Which ancient civilization is closely connected with linen fabric?

A1: When the tombs of Egypt's pharaohs were opened and the mummies discovered, archaeologists found that they had been wrapped in swathes of linen, and both the bodies and the fabric were found to be well preserved. This reveals that linen had been around for more than 6 millennia and that it was a fiber being widely used by humans throughout the ages.

Q2: Name the countries that are the largest producers of linen fiber.

A2: The largest producers of this plant are the European nations of France, Belgium, and the Netherlands.

Q3: In what respect does linen score over its close competitor: cotton?

A3: Linen is at least three times stronger than cotton.

Q4: Linen is used to make space suits because of a certain characteristic unique to the fiber. Name the characteristic.

A4: Linen's exceptional capacity to resist static electric load makes it a preferred choice for use in space suits.

Chapter 19 | **Organic Wool** *Lesson Nineteen*

Introduction – Three Things You Know
- Have students write down three things they know about organic wool.
- Have them share their knowledge with the class and make a list on the board.
- Try to answer any questions they have and facilitate any discussion that happens.
- Tell the students about the assigned reading and that you are going to show them a Power Point presentation and that they must write down three things they learn from both of these exercises. Have them save their answers to share later.

Reading
- Allow reading time for the organic wool chapter in class.
- Have students finish reading the organic wool chapter at home.

Power Point Presentation
- Show the class the organic wool chapter Power Point presentation.
- Talk through each slide and explain the contents.
- Students are to write down three things they learned.
- Answer any questions as best you can and foster discussion.

Celebrity Profile
- Using the internet and other resources, have students choose an eco-minded celebrity to profile in a short, yet complete, one-page report. Instruct students to include the reasons why the celebrity is eco-minded and have them list some of the celebrity's associations to the sustainability community.

Quick Answer: Sustainability Approach
- During class, ask students the following question in an open discussion format: What would be your approach to developing an innovative solution to the following: organic livestock?

Three Things You've Learned
- Students will review their original organic wool-related entries and compare them to what they've learned during the reading and Power Point exercises.

Organic Wool Chapter Quiz
- Students will take the organic wool chapter quiz located within the Student Workbook.

Organic Wool Chapter Quiz

Q1: How does organic wool differ from conventional wool?

A1: The use of conventional wool is slowly losing pace to organic wool in man's race to show himself as an ecologically sensitive creature. Organic wools are free from chemicals, such as pesticides, as they come from sheep which are not exposed to toxins.

Q2: Which countries are the main producers of organic wool?

A2: Most organic wool comes from New Zealand and Australia, the prime producers of almost every kind of wool. Organic wool can also be sourced from England, the United States, Turkey, and Argentina, among other countries.

Q3: What unique qualities of organic wool are responsible for its strength and durability?

A3: The crimping property of organic wool makes it naturally resilient and elastic, as the fiber bends and turns into a three-dimensional structure. This property makes the fiber durable, warm, and resistant to abrasion. This fiber can also absorb 30% of its weight in moisture which makes it suitable for all kinds of climates. Since wool retains moisture it resists flames, and upon burning, chars and extinguishes naturally. The durability of organic wool is such that it can be bent 20,000 times and not break.

Q4: What prohibitive aspects of sheep rearing ensure the eco-quality of organic wool?

A4: These sheep have been reared in absolutely healthy conditions, fit for both survival and for production of high quality wool. Produced from healthy varieties of sheep raised on specialized organic farms, the wool produced is soft and comfortable.

Chapter 20 | Piña

Lesson Twenty

Introduction – Three Things You Know

- Have students write down three things they know about piña.
- Have them share their knowledge with the class and make a list on the board.
- Try to answer any questions they have and facilitate any discussion that happens.
- Tell the students about the assigned reading and that you are going to show them a Power Point presentation and that they must write down three things they learn from both of these exercises. Have them save their answers to share later.

Reading

- Allow reading time for the piña chapter in class.
- Have students finish reading the piña chapter at home.

Power Point Presentation

- Show the class the piña chapter Power Point presentation.
- Talk through each slide and explain the contents.
- Students are to write down three things they learned.
- Answer any questions as best you can and foster discussion.

Social/ Mobile

- During class, ask students the following question in an open discussion format: Discuss the role of social media and mobile technology in engaging the sustainability community and spreading awareness among the young generation about green fabrics and green design. How is it helping to create an informed generation that understands the values and benefits of green fabrics?

Quick Answer: Sustainability Approach

- During class, ask students the following question in an open discussion format: What would be your approach to developing an innovative solution to the following: organic farming equipment?

Three Things You've Learned

- Students will review their original piña-related entries and compare them to what they've learned during the reading and Power Point exercises.

Piña Chapter Quiz

- Students will take the piña chapter quiz located within the Student Workbook.

Piña Chapter Quiz

Q1: Name the plant from which piña fiber is extracted.

A1: Piña is made from the pineapple palm plant, which is also referred to as *piña.*

Q2: Which countries are primarily associated with the cultivation of pineapples?

A2: Pineapple is widely cultivated in the Philippines, Hawaii, India, West Indies, Indonesia, South America, East Asia, Sri Lanka, Africa, and Australia.

Q3: What other environmentally-friendly plants are closely associated with the production techniques of piña?

A3: Bamboo and coconut shells, which are used to expose the epidermis, which leaves the fiber revealed for extraction, are the other environmentally-friendly plants closely associated with the production of piña fiber.

Q4: How is filament yarn created with piña and why does it lead to the presence of knots in the finished fabric?

A4: To create a filament effect, the fibers are knotted at the ends with one another, after which they are transferred to a spinning wheel. Weaving a fabric without knots is almost impossible as the fibers are extremely susceptible to breakage and need the knots for reducing breakage.

Chapter 21 | Ramie

Lesson Twenty-One

Introduction – Three Things You Know

- Have students write down three things they know about ramie.
- Have them share their knowledge with the class and make a list on the board.
- Try to answer any questions they have and facilitate any discussion that happens.
- Tell the students about the assigned reading and that you are going to show them a Power Point presentation and that they must write down three things they learn from both of these exercises. Have them save their answers to share later.

Reading

- Allow reading time for the ramie chapter in class.
- Have students finish reading the ramie chapter at home.

Power Point Presentation

- Show the class the ramie chapter Power Point presentation.
- Talk through each slide and explain the contents.
- Students are to write down three things they learned.
- Answer any questions as best you can and foster discussion.

News Article Review

- Have students research a current news article on eco-friendly fabrics. Ask students to describe its contents in detail and provide insights to the class about its relevancy to sustainability.

Quick Answer: Sustainability Approach

- During class, ask students the following question in an open discussion format: What would be your approach to developing an innovative solution to the following: non-mechanical spinning and weaving methods?

Three Things You've Learned

- Students will review their original ramie-related entries and compare them to what they've learned during the reading and Power Point exercises.

Ramie Chapter Quiz

- Students will take the ramie chapter quiz located within the Student Workbook.

Ramie Chapter Quiz

Q1: Elaborate on the affection shown toward ramie by the Chinese. Who was presented with gifts of ramie as a sign of respect and gratitude?

A1: The Chinese are closely associated with and absolutely prize ramie, and in times past, garments made of this translucent fabric were brought out of treasure chests to celebrate the onset of summer. A specific day, the 5th day of the 5th lunar month according to the Chinese calendar, was set out for festivities. Young women would deck themselves in their finest ramie clothes and cavort, unbridled by the bulk of their winter garments. Chinese emperors were presented with gifts of ramie as a mark of respect and gratitude.

Q2: Which special properties of ramie made it ideal as a wrapping for mummies?

A2: A fiber of Chinese origin, ramie has been found in the tombs of the pharaohs of Egypt, testifying to the special status of the fabric that comes from its innate mold and bacteria resistant qualities.

Q3: What makes ramie fabric more susceptible to brittleness compared to other natural fibers? How does it rate in terms of resilience, elasticity, and elongation?

A3: Because of its high molecular crystallinity, ramie is more brittle and less durable than other natural fibers. Therefore, it is blended with various fibers so as to enhance its strength while imparting a silky texture to the blend. Ramie shows good absorption and density, but when it comes to resilience, elasticity, and elongation, it is found lacking. It's for these reasons that the fabric tends to be brittle and also breaks and tears if it is repeatedly folded at the same point.

Q4: At what stage of the plant's growth is ramie harvested? Where is it best to cut the stem?

A4: The harvesting is done either before or after the plant begins to flower. This is the best suited time for maximum fiber extraction with the stem of the plant being cut close to the lateral root.

Chapter 22 | Recycled Nylon

Lesson Twenty-Two

Introduction – Three Things You Know

- Have students write down three things they know about recycled nylon.
- Have them share their knowledge with the class and make a list on the board.
- Try to answer any questions they have and facilitate any discussion that happens.
- Tell the students about the assigned reading and that you are going to show them a Power Point presentation and that they must write down three things they learn from both of these exercises. Have them save their answers to share later.

Reading

- Allow reading time for the recycled nylon chapter in class.
- Have students finish reading the recycled nylon chapter at home.

Power Point Presentation

- Show the class the recycled nylon chapter Power Point presentation.
- Talk through each slide and explain the contents.
- Students are to write down three things they learned.
- Answer any questions as best you can and foster discussion.

My Own Way

- Ask students if they have ever been in a situation whereby they were confronted with challenges to their eco-friendly beliefs. What did they do and how did they and others react? This assignment can be discussed in an open format with the entire class.

Quick Answer: Sustainability Approach

- During class, ask students the following question in an open discussion format: What would be your approach to developing an innovative solution to the following: local farming and economic development?

Three Things You've Learned

- Students will review their original recycled nylon-related entries and compare them to what they've learned during the reading and Power Point exercises.

Recycled Nylon Chapter Quiz

- Students will take the recycled nylon chapter quiz located within the Student Workbook.

Recycled Nylon Chapter Quiz

Q1: What is pre-consumer waste and how is it associated with recycled nylon fibers?

A1: The image of landfills bulging with discarded plastic bottles, waste cuttings from industrial products, and pre-consumer waste has become a part of our global landscape. Recycled nylon uses waste that would otherwise have taken decades to decompose and converts it into beautiful, green fiber. There is also pre-consumer waste generated during production. The use of these waste materials provides the resources needed to manufacture recycled nylon.

Q2: Name one product that uses recycled nylon fibers for production with the product itself being recycled for producing new recycled nylon fiber.

A2: The ubiquitous office carpet, whose primary function is to be durable, is one of the most widespread applications of recycled nylon. It provides the raw material needed to produce it again, upon discarding it after much wear and tear. With even a fraction of the 250 million pounds being recycled annually, it creates a huge volume of recycled nylon.

Q3: Sources must be accessed for raw material needed for recycled nylon production. Which two nations are the most significant for this purpose?

A3: Areas where significant industrial and consumer waste are generated provide the resources needed to manufacture recycled nylon. On that basis, China and the United States of America are the two nations that have seen primacy in its production.

Q4: Describe the various steps, from accessing raw material to the finished fiber, that are used in producing recycled nylon.

A4: The production of recycled nylon begins with accumulating available pre and post-industrial waste, and then converting it into chips, which are melted into polymer. To make these chips, the raw material is chopped, ground, and melted, and then put through a conversion process, which creates nylon chips through reformulation of the material. It is these chips that are turned into filaments using a proprietary process unique to individual companies, although they follow a basic procedure to create the polymer. This polymer is then extruded through spinnerets. The resultant filament is combined together to make strong yarn, which is processed in a way that allows it to acquire bulk, strength, and crimp. The process of texturing is what gives the fabric its elasticity, comfort, and softness.

Chapter 23 | Recycled Polyester

Lesson Twenty-Three

Introduction – Three Things You Know

- Have students write down three things they know about recycled polyester.
- Have them share their knowledge with the class and make a list on the board.
- Try to answer any questions they have and facilitate any discussion that happens.
- Tell the students about the assigned reading and that you are going to show them a Power Point presentation and that they must write down three things they learn from both of these exercises. Have them save their answers to share later.

Reading

- Allow reading time for the recycled polyester chapter in class.
- Have students finish reading the recycled polyester chapter at home.

Power Point Presentation

- Show the class the recycled polyester chapter Power Point presentation.
- Talk through each slide and explain the contents.
- Students are to write down three things they learned.
- Answer any questions as best you can and foster discussion.

Essay

- Have students write an 800-word essay on the collective excitement presently seen in the design industry as indicated by the number of eco-fashion and interiors conferences and expositions being held on an international scale. Which kinds of attendees do these conferences attract? What features, speakers, and workshops are being presented to solidify the idea of green design? Remind students to use cited sources in their responses.

Three Things You've Learned

- Students will review their original recycled polyester-related entries and compare them to what they've learned during the reading and Power Point exercises.

Recycled Polyester Chapter Quiz

- Students will take the recycled polyester chapter quiz located within the Student Workbook.

Recycled Polyester Chapter Quiz

Q1: Name some of the most important qualities of synthetic fibers seen in garments made from recycled polyester.

A1: One of the most important qualities of synthetic fibers seen in garments made from recycled polyester come from its synthetic nature which discourages growth of bacteria and microbes leading to a mold and mildew free fabric, even when wet. In addition, its anti-wrinkle nature does not exhibit any shrinkage, and it has high strength and durability. Being a filament yarn, protruding ends are eliminated leading to lint-free and pill-free fabrics.

Q2: What material is used to create recycled polyester?

A2: Existing, traditional polyester (polyethylene terephthalate (PET)) is used to develop recycled polyester.

Q3: What is *antimony* and how does it harm the environment?

A3: *Antimony*, a toxic chemical compound that is similar to compounds found in arsenic, is converted to antimony trioxide when it reaches high temperatures. The recycling process releases all of the fabric's compounds, including carcinogens, therefore allowing them to be taken in by humans, animals, and the land. This can cause breathing and respiratory problems including emphysema and other concerns. Further, toxic wastewater is the result of antimony trioxide being washed into it, and possibly into waterways.

Q4: How many barrels of oil are used annually to produce original virgin polyester?

A4: Facts indicate that the estimation of the amount of barrels of oil it takes to produce traditional, virgin polyester is quite staggering at 70 million annually.

Chapter 24 | **Recycled Wool** *Lesson Twenty-Four*

Introduction – Three Things You Know

- Have students write down three things they know about recycled wool.
- Have them share their knowledge with the class and make a list on the board.
- Try to answer any questions they have and facilitate any discussion that happens.
- Tell the students about the assigned reading and that you are going to show them a Power Point presentation and that they must write down three things they learn from both of these exercises. Have them save their answers to share later.

Reading

- Allow reading time for the recycled wool chapter in class.
- Have students finish reading the recycled wool chapter at home.

Power Point Presentation

- Show the class the recycled wool chapter Power Point presentation.
- Talk through each slide and explain the contents.
- Students are to write down three things they learned.
- Answer any questions as best you can and foster discussion.

Open Discussion

- Discuss the following with the class in an open format: Describe your version of ideal government support systems for green designers. What would this look like and what incentives might there be? Is there a role for the government in green design?

Quick Answer: Sustainability Approach

- During class, ask students the following question in an open discussion format: What would be your approach to developing an innovative solution to the following: environmentally-friendly shipping methods?

Three Things You've Learned

- Students will review their original recycled wool-related entries and compare them to what they've learned during the reading and Power Point exercises.

Recycled Wool Chapter Quiz

- Students will take the recycled wool chapter quiz located within the Student Workbook.

Recycled Wool Chapter Quiz

Q1: How is wool produced?

A1: Wool is a fiber that is produced by the follicles of certain members of the Caprinae family, to which sheep belong. Goats, llamas and rabbits also provide humans with wool, but not of the same quality as found in the sheep family.

Q2: What is the difference between the hairy fiber of sheep and fibers obtained from other animals?

A2: Sheep's wool is set apart from the hairy fibers taken from other animals by the ease with which it can be spun. The ease in spinning results from the scaling and crimp of the wool, which allows for easy attachment of one fiber to another.

Q3: Describe *crimp* and how it aids in spinning.

A3: The *crimp* of the wool is the feature which allows for easy attachment of one fiber to another. It is the crimp, or the inherent waviness of wool, that allows for air retention and bulk in the fiber, endowing it with insulation.

Q4: What quality of wool makes it the favorite fiber of the Bedouins?

A4: The most amazing quality of its insulation is the ability to trap heat in cold climates and to keep the heat out in warmer places. This makes it the preferred choice for the nomads of the desert, the Bedouins.

Chapter 25 | **Rush**
Lesson Twenty-Five

Introduction – Three Things You Know

- Have students write down three things they know about rush.
- Have them share their knowledge with the class and make a list on the board.
- Try to answer any questions they have and facilitate any discussion that happens.
- Tell the students about the assigned reading and that you are going to show them a Power Point presentation and that they must write down three things they learn from both of these exercises. Have them save their answers to share later.

Reading

- Allow reading time for the rush chapter in class.
- Have students finish reading the rush chapter at home.

Power Point Presentation

- Show the class the rush chapter Power Point presentation.
- Talk through each slide and explain the contents.
- Students are to write down three things they learned.
- Answer any questions as best you can and foster discussion.

Social/ Mobile

- During class, ask students the following question in an open discussion format: Describe how social media and mobile technology can assist up and coming green designers with their business practices. Does social media allow for instant promotion of eco-brands, and does the public respond to these messages? How can mobile technology help with the organization of one's green design business?

Quick Answer: Sustainability Approach

- During class, ask students the following question in an open discussion format: What would be your approach to developing an innovative solution to the following: sustainable harvesting methods?

Three Things You've Learned

- Students will review their original rush-related entries and compare them to what they've learned during the reading and Power Point exercises.

Rush Chapter Quiz

- Students will take the rush chapter quiz located within the Student Workbook.

Rush Chapter Quiz

Q1: What is *tatami*?

A1: *Tatami* is the Japanese word for floor mats made of soft rushes.

Q2: Why are items made of rush extremely lightweight?

A2: Items made of rush are extremely lightweight because of the hollow stems of the rushes.

Q3: When is the best time to gather rushes for the purpose of making fibers?

A3: Rush should be gathered when the leaves are green, or only when the tips of the leaves and stalks are beginning to turn brown. This period corresponds to late July through early September.

Q4: Why is it important to ensure that harvested rush plans are kept in the dark and kept damp?

A4: The drying process should be undertaken in dark, airy rooms after making bundles of similarly sized leaves. This is done so as to avoid excess sunlight, which may make the leaves brittle. Rush should always be dampened with water, as wetting makes the leaves pliable and easy to work with because dry rush tends to break easily. The resultant, woven products are quite stunning in visual quality and are imported and exported worldwide.

Chapter 26 | Sea Cell

Lesson Twenty-Six

Introduction – Three Things You Know

- Have students write down three things they know about sea cell.
- Have them share their knowledge with the class and make a list on the board.
- Try to answer any questions they have and facilitate any discussion that happens.
- Tell the students about the assigned reading and that you are going to show them a Power Point presentation and that they must write down three things they learn from both of these exercises. Have them save their answers to share later.

Reading

- Allow reading time for the sea cell chapter in class.
- Have students finish reading the sea cell chapter at home.

Power Point Presentation

- Show the class the sea cell chapter Power Point presentation.
- Talk through each slide and explain the contents.
- Students are to write down three things they learned.
- Answer any questions as best you can and foster discussion.

Creative Worksheet

- Textiles. Have students research and present a short oral report on the use of a specific green fiber providing detail on its origins, properties, and characteristics. Have students explain its weave, thickness, and durability, and discuss its availability in the marketplace.

Three Things You've Learned

- Students will review their original sea cell-related entries and compare them to what they've learned during the reading and Power Point exercises.

Sea Cell Chapter Quiz

- Students will take the sea cell chapter quiz located within the Student Workbook.

Sea Cell Chapter Quiz

Q1: What is the raw material for making sea cell fabric and what process is used in its manufacture?

A1: The unfathomable depths of the seas and oceans hide within them unknown treasures, and one among these is found in seaweed, which can be transformed into fibers for making fabrics using modern technology. Seaweed fibers cannot stand on their own, and so a cellulose-based fabric is created using the lyocell process and it is this fabric which serves as the basis for sea cell fabric, commercially known as *seaweed.*

Q2: How does the body absorb nutrients released by sea cell fabric?

A2: Seaweed is full of nutrients and by transforming seaweed into fabric humans are able to partake of these nutrients, which nourish the skin. Moisture released by the body facilitates the absorption of calcium, magnesium, vitamins, and amino acids, which occur naturally in seaweed.

Q3: Which garments are most often created using sea cell fibers?

A3: Sea cell is mainly used to manufacture items where hygiene is at a premium, such as undergarments, personal hygiene products, and household products including bed sheets, towels, and carpets.

Q4: How is seaweed used by different cultures?

A4: Different cultures use seaweed for different purposes. While the Japanese and Chinese prize seaweed for its therapeutic and curative properties, people in Britain and Ireland have been known to value seaweed as a nutrient rich food and it is used as a dietary supplement.

Chapter 27 | Sea Grass
Lesson Twenty-Seven

Introduction – Three Things You Know

- Have students write down three things they know about sea grass.
- Have them share their knowledge with the class and make a list on the board.
- Try to answer any questions they have and facilitate any discussion that happens.
- Tell the students about the assigned reading and that you are going to show them a Power Point presentation and that they must write down three things they learn from both of these exercises. Have them save their answers to share later.

Reading

- Allow reading time for the sea grass chapter in class.
- Have students finish reading the sea grass chapter at home.

Power Point Presentation

- Show the class the sea grass chapter Power Point presentation.
- Talk through each slide and explain the contents.
- Students are to write down three things they learned.
- Answer any questions as best you can and foster discussion.

Mass Media

- Have students break into groups to discuss how green fabrics and other environmental concerns are portrayed in the mass media. Assign each group the task of bringing to class an article, print ad, online promotion, film review, or advertisement showing how green fabrics, green products, and related concerns are promoted or portrayed.

Three Things You've Learned

- Students will review their original sea grass-related entries and compare them to what they've learned during the reading and Power Point exercises.

Sea Grass Chapter Quiz

- Students will take the sea grass chapter quiz located within the Student Workbook.

Sea Grass Chapter Quiz

Q1: What is sea grass and where is it found?

A1: Long, narrow, leafy marine flowering plants that grow in meadows like formations in the sea are labeled *sea grass* and the fibers derived are used to create some of the finest and most stylish home accessories in the world. It is a wholly organic, biodegradable, naturally occurring resource, which can be used for a variety of purposes, especially in the home furnishings sector. It is grown in paddy fields, which are flooded with sea water.

Q2: Which countries are the chief cultivators of sea grass?

A2: Sea grass is primarily grown in South East Asian nations with China, Vietnam, and India leading the supply of fibers.

Q3: In what ways is sea grass an environment protector apart from its use as a green fiber?

A3: Sea grass is a wonderful friend of the environment as it engineers the creation of its own habitat in the seas. Its leaves control the flow of sediments that travel through the rivers, and its own roots and rhizomes hold the seabed together. Creating a natural impediment to the harmful effects of fast moving sea currents, sea grass is also an important shelter for several small plant and animal species.

Q4: How does sea grass get its unique look?

A4: The fiber is allowed to retain certain impurities, and these impart each product with a unique look. Sea grass is one of the most environmentally-friendly of all organic materials because it is generally allowed to retain its natural green color, thus doing away with the dyeing process altogether.

Chapter 28 | **Silk**
Lesson Twenty-Eight

Introduction – Three Things You Know

- Have students write down three things they know about silk.
- Have them share their knowledge with the class and make a list on the board.
- Try to answer any questions they have and facilitate any discussion that happens.
- Tell the students about the assigned reading and that you are going to show them a Power Point presentation and that they must write down three things they learn from both of these exercises. Have them save their answers to share later.

Reading

- Allow reading time for the silk chapter in class.
- Have students finish reading the silk chapter at home.

Power Point Presentation

- Show the class the silk chapter Power Point presentation.
- Talk through each slide and explain the contents.
- Students are to write down three things they learned.
- Answer any questions as best you can and foster discussion.

Essay

- Have students write an 800-word essay on what it would take for a designer to redirect their career from being a traditional designer to a green designer. What kinds of societal or emotional occurrences might take place that would cause such a thing to happen? What kind of industry exposure and education would they need to have in order to be effective in this new career choice? Would it be important for their personal values to be promoted through their brand? Ask students to explain their insights in detail and remind them to use cited sources in their responses.

Three Things You've Learned

- Students will review their original silk-related entries and compare them to what they've learned during the reading and Power Point exercises.

Silk Chapter Quiz

- Students will take the silk chapter quiz located within the Student Workbook.

Silk Chapter Quiz

Q1: Explain the connection between silk and tea.

A1: A Chinese princess was said to be sipping tea in her garden when a cocoon, dangling from a branch, fell into her hot tea. It suddenly released threads of silk, which later became the chosen fiber of the Imperial household.

Q2: What is the Silk Route?

A2: The trade of silk with Europe and the Near East led to the establishment of the fabled Silk Route, used chiefly for trading in silk.

Q3: Name some of the historical personalities that showed affinity for silk fabrics.

A3: Aristotle, Alexander, the ancient Greeks and Romans, the fathers of the early church, and European nobility were admirers of this lustrous, free draping fabric leading to the universalization of silk fabric.

Q4: How did silk making reach the rest of the world from China?

A4: The Chinese zealously guarded the art of making silk, but like all coveted arts this too was smuggled out of the country to nearby Japan in 300 A.D. from where it slowly made its way to India, Spain, and Italy.

Chapter 29 | Soybean Protein

Lesson Twenty-Nine

Introduction – Three Things You Know

- Have students write down three things they know about soybean protein.
- Have them share their knowledge with the class and make a list on the board.
- Try to answer any questions they have and facilitate any discussion that happens.
- Tell the students about the assigned reading and that you are going to show them a Power Point presentation and that they must write down three things they learn from both of these exercises. Have them save their answers to share later.

Reading

- Allow reading time for the soybean protein chapter in class.
- Have students finish reading the soybean protein chapter at home.

Power Point Presentation

- Show the class the soybean protein chapter Power Point presentation.
- Talk through each slide and explain the contents.
- Students are to write down three things they learned.
- Answer any questions as best you can and foster discussion.

Celebrity Profile

- Using the internet and other resources, have students choose an eco-minded celebrity to profile in a short, yet complete, one-page report. Instruct students to include the reasons why the celebrity is eco-minded and have them list some of the celebrity's associations to the sustainability community.

Three Things You've Learned

- Students will review their original soybean protein-related entries and compare them to what they've learned during the reading and Power Point exercises.

Soybean Protein Chapter Quiz

- Students will take the soybean protein chapter quiz located within the Student Workbook.

Soybean Protein Chapter Quiz

Q1: What are the reasons for the skin enhancing properties of soybean protein?

A1: Soybean Protein Fiber (SPF) has 16 amino acids that promote healthy skin invisibly while nourishing damaged skin.

Q2: Why is soybean fiber also referred to as vegetable cashmere?

A2: Soybean is the only botanic protein fiber in the world, often referred to as *vegetable cashmere* for its smooth, soft appeal.

Q3: Which famous American was among the first to use soybean fiber, for both his garments and automobile?

A3: Henry Ford.

Q4: Where was soybean protein fiber traditionally cultivated?

A4: Soybean protein fiber is cultivated in almost all parts of the globe, although it was traditionally an East Asian crop grown in China, Japan, and Korea.

Chapter 30 | **Spider Silk** *Lesson Thirty*

Introduction – Three Things You Know
- Have students write down three things they know about spider silk.
- Have them share their knowledge with the class and make a list on the board.
- Try to answer any questions they have and facilitate any discussion that happens.
- Tell the students about the assigned reading and that you are going to show them a Power Point presentation and that they must write down three things they learn from both of these exercises. Have them save their answers to share later.

Reading
- Allow reading time for the spider silk chapter in class.
- Have students finish reading the spider silk chapter at home.

Power Point Presentation
- Show the class the spider silk chapter Power Point presentation.
- Talk through each slide and explain the contents.
- Students are to write down three things they learned.
- Answer any questions as best you can and foster discussion.

Open Discussion
- Discuss the following with the class in an open format: What are the advantages of sustainable fabric production methods? Consider energy utilization, locality, transportation, local resources, labor, and costs. What are the disadvantages?

Three Things You've Learned
- Students will review their original spider silk-related entries and compare them to what they've learned during the reading and Power Point exercises.

Spider Silk Chapter Quiz
- Students will take the spider silk chapter quiz located within the Student Workbook.

Spider Silk Chapter Quiz

Q1: What is another name for spider silk?

A1: Gossamer.

Q2: Why is spider silk called the *holy grail of fibers*?

A2: Humans have used spider silk for thousands of years, and the limited availability combined with its extraordinary applications has earned it the nickname *holy grail of fibers*.

Q3: What quality of spider silk made it especially valuable for people in ancient times?

A3: In ancient times, spider silk was valued more for its healing properties than as a clothing fiber.

Q4: Who pioneered the process of spider silk cultivation?

A4: A French missionary, Jacob Paul Cambone, living in Madagascar in the 1880s took to the task of producing silk from spiders and engineered a machine that would employ 24 spiders at a single time.

Chapter 31 | **Tencel® / Lyocell** *Lesson Thirty-One*

Introduction – Three Things You Know
- Have students write down three things they know about Tencel®/ lyocell.
- Have them share their knowledge with the class and make a list on the board.
- Try to answer any questions they have and facilitate any discussion that happens.
- Tell the students about the assigned reading and that you are going to show them a Power Point presentation and that they must write down three things they learn from both of these exercises. Have them save their answers to share later.

Reading
- Allow reading time for the Tencel®/ lyocell chapter in class.
- Have students finish reading the Tencel®/ lyocell chapter at home.

Power Point Presentation
- Show the class the Tencel®/ lyocell chapter Power Point presentation.
- Talk through each slide and explain the contents.
- Students are to write down three things they learned.
- Answer any questions as best you can and foster discussion.

My Own Way
- Ask students if they have ever been in a situation whereby they were confronted with challenges to their eco-friendly beliefs. What did they do and how did they and others react? This assignment can be discussed in an open format with the entire class.

Three Things You've Learned
- Students will review their original Tencel®/ lyocell-related entries and compare them to what they've learned during the reading and Power Point exercises.

Tencel® / Lyocell Chapter Quiz
- Students will take the Tencel®/ lyocell chapter quiz located within the Student Workbook.

Tencel®/ Lyocell Chapter Quiz

Q1: What are some of the characteristics of Tencel® and how does it relate to lyocell?

A1: Tencel® is the trade name for lyocell, a man-made natural fiber. It has an extremely flattering drape that clings sensuously to the human form, sending out a signal of luxurious elegance. It is soft, breathable, lightweight, comfortable, and shrink resistant. It has good strength not only when wet, but also when dry.

Q2: Why do colors take on a rich look with Tencel®?

A2: Tencel® dyes uniformly and all colors have the propensity to acquire a richer look when applied to this fabric.

Q3: What quality does Tencel® impart to synthetic fibers?

A3: When it is blended with synthetic fibers it adds a dimension of shape retention to the fabric.

Q4: Tencel® fiber is a sub-species of which synthetic fiber?

A4: The fiber is considered a sub-species of rayon.

Chapter 32 | Wild Nettle

Lesson Thirty-Two

Introduction – Three Things You Know

- Have students write down three things they know about wild nettle.
- Have them share their knowledge with the class and make a list on the board.
- Try to answer any questions they have and facilitate any discussion that happens.
- Tell the students about the assigned reading and that you are going to show them a Power Point presentation and that they must write down three things they learn from both of these exercises. Have them save their answers to share later.

Reading

- Allow reading time for the wild nettle chapter in class.
- Have students finish reading the wild nettle chapter at home.

Power Point Presentation

- Show the class the wild nettle chapter Power Point presentation.
- Talk through each slide and explain the contents.
- Students are to write down three things they learned.
- Answer any questions as best you can and foster discussion.

News Article Review

- Have students research a current news article on eco-friendly fabrics. Ask students to describe its contents in detail and provide insights to the class about its relevancy to sustainability.

Three Things You've Learned

- Students will review their original wild nettle-related entries and compare them to what they've learned during the reading and Power Point exercises.

Wild Nettle Chapter Quiz

- Students will take the wild nettle chapter quiz located within the Student Workbook.

Wild Nettle Chapter Quiz

Q1: Which country is notably associated with wild nettle plant cultivation?

A1: The Himalayan nation of Nepal.

Q2: Which ancient tribe used wild nettle to make their fishing nets?

A2: Natives of the Algonquin tribe were known to make fishing nets from this fiber.

Q3: What is the connection between German soldiers and wild nettle?

A3: During the World Wars in the absence of adequate stocks of cotton, the German army turned towards making uniforms for their soldiers with fabric made from wild nettle.

Q4: What conditions are needed for growth of the wild nettle plant?

A4: Wild nettle is found in almost all parts of the globe that enjoy a temperate climate. Being a weed, it spreads quickly without excess human input, and any soil that has been properly tended is the perfect breeding ground for this plant.

Lesson Plans

Teaching Companion
Guide to Green Fabrics

Section 4

Answer Key
Final Exam – Extra Credit # 6 - 19

Answer Key

Final Exam Questions – Part I
Guide to Green Fabrics

1. List five eco-friendly characteristics of the Abaca plant.

Examples: 100 % biodegradable, promotes biodiversity, naturally colored from ivory to brown, prevents soil erosion, sustainable, regenerates itself naturally through suckering, provides weed suppression.

2. Can bamboo truly be called an eco-friendly fabric? List reasons why you agree or disagree.

Yes, bamboo can truly be referred to as an eco-friendly fabric because of the following reasons: it is biodegradable, it enriches the soil, and it grows without need for pesticides or fertilizers.

3. Relative to gold and diamonds Cervelt™ is considered more expensive. Why?

Cervelt™ is considered more expensive relative to gold and diamonds since it is 200 times rarer than gold, 3,000 times rarer than cashmere, and nearly 2 ½ times rarer than diamonds because of the scarcity of the raw material.

4. Why are more outdoor furnishings and household items made of coir than indoor furnishings?

Being immune to dampness and moisture, Coir products are popular in fabricating outdoor items, which are prone to the vagaries of the weather, as it retains its inherent properties even in adverse weather conditions.

5. What two important human needs does corn cultivation satisfy?

Corn satisfies the dietary needs as well as fiber concerns of ever increasing populations, making it one of the greenest fibers on earth.

6. What feature of the EcoFi® yarn is responsible for keeping it pill-free and lint-free?

Due to the significant length of the yarn, protruding ends are eliminated leading to lint-free and pill-free fabrics.

7. Describe organic leather.

Organically raised animals that graze freely, roam at their leisure, and generally live under stress-free conditions produce hides that generally classify as organic. Should their hides be tanned organically as well, then the resultant leather would classify as organic leather. Both conditions must be present for this classification to apply. Because this type of leather is not yet very common, organic leather is out of reach for some. Products made from this leather include bags, belts and furniture.

8. Name several byproducts of hemp, and their uses.

By-products of harvested hemp are also useful in other applications. The core fiber can be used to make paper, mulch, horse bedding, litter, and construction materials. Hemp dust, created during the production process, is pressed into pellets and used for fuel, and small chips from the core are used as a high nutrient soil additive.

9. What are some of the most important uses of henequen, especially in the automobile industry?

As a reinforcement material, it is used in composite plastic manufacture, especially in auto components. Brake pads use henequen as a substitute for asbestos.

10. Ingeo fiber combines the qualities of three natural fibers. Name these three fibers.

Its characteristics are akin to silk, wool, and cotton, three natural fibers, in touch, feel and texture.

11. Explain the steps that should be used in the care of products made from jute.

Jute rugs can be easily vacuumed, and in case of spills, it is best to first blot and then clean the spot with a wet cloth. To take care of mud stains, they are best cleaned after they have been allowed to dry, then scraped away, and finally cleaned with a wet cloth. Use of acids for cleaning should be avoided as it reduces the strength of the fiber.

12. How is blended kenaf fabric cared for?

Since most of its clothing related applications involve these fibers being blended with cotton fibers, the fabric will require the same kind of care as cotton fibers, as well as linen. Items may be cleaned in a traditional washing machine or dry-cleaned.

13. Name five qualities of milk protein fiber.

Options include: anti-bacterial, anti-fungal, has 18 amino acids that nourish the skin, has good moisture retention, has good wicking abilities, and has natural, wet protecting genes.

14. What precaution must be taken in caring for modal fabric?

Fabrics that are made of 100% modal need to be lightly ironed after washing, but apart from this there are no other major care concerns. Even when washed in warm water it retains its vibrant color, but low temperature washing is recommended. It has sensitivity to chlorine bleach, which should be used in minimal amounts. The fabric should be dried according to directions, which likely will be similar to those for drying cotton.

15. What important precautions must be strictly followed when taking care of garments and items made from latex? How should these items be stored?

Caring for latex products is not time consuming, but certain precautions and procedures need to be followed. Garments made from latex should be rinsed clean in warm water and then hung upside down to be dried. Use of solvents is not advised as the chemicals contained in them may lead to breakdown of the chemical composition of natural latex and render the garment useless. The fiber also lacks resistance to oxidizing agents and can be damaged by oil, sunlight, perspiration, and aging. Therefore, natural latex garments should be stored in a cool, dry place away from sunlight.

16. What scientific innovation has led to the elimination of the dyeing process in the production of some organic cotton?

With scientific research and application of newer ways of cultivation, naturally colored organic cotton has emerged and is grown on the stem, eliminating the need for the dyeing process completely. However, the range of colors at present is restricted to brown, reddish-brown, green, and yellow. Unlike dyed cotton, which fades with washing, naturally colored organically grown cotton exhibits a propensity towards deepening of color during its first three washes.

17. Explain the method used for washing clothes and products made of organic wool.

Washing should be done using a biodegradable detergent. This fabric does not take well to alkaline detergents. The water temperature should be mild. The preferred mode for washing is hand washing. After removing an organic wool garment from the washing tub, gently shake it to remove excess water. The garment should be placed on a dry cotton towel and laid flat. This helps to remove moisture and odor. Organic wool should not be dried in direct sunlight as it may result in color fading. Once the garment is completely dry it should be folded flat and wrapped in a natural wrapper, but never in plastic.

18. List five qualities of piña fiber.

Options include: delicate yet very strong, durable, does not easily wear, has excellent textural qualities, and is lightweight with an elegant, beautiful appearance.

19. How is gum removed from ramie fiber?

Unlike other bast fibers, this fiber has to be chemically processed to remove the gum, which is present in the stem.

20. In terms of its material base, how is recycled polyester a green fabric?

Recycled polyester is simply a recycled version of polyester, and an environmentally-friendly product with far reach, depending on its composition and methods. The raw materials used for making recycled polyester are abundantly available, and by using discarded materials, the recycling of these fibers aids in protecting the environment, which benefits the ecosystem. In addition, manufacturing recycled polyester reduces waste production and petroleum consumption.

21. Why are moths attracted to woolen garments, and how can they be repelled?

Moths are attracted to wool because of its natural sulfur content. Protection from moths and their larvae should be done by placing cedar blocks, moth balls, or herbal sachets near garments.

22. Give another name for rush.

Fiber rush.

23. From where is seaweed sourced?

Seas and oceans abound with seaweed, which is a naturally occurring renewable resource.

24. What qualities of sea grass make it a 100% green fiber?

Sea grass is a completely green fiber as there are no chemical inputs in growing, harvesting, and fabricating the fiber, and also because it is 100% biodegradable. It holds much promise to be one of the most important fibers of the future.

25. Why is the acid dye method the best way to dye silk fibers?

Silk is best dyed using the acid dye method, where the dye is incorporated into the fiber at a cellular level.

Answer Key | **Final Exam Term Definitions – Part II** *Guide to Green Fabrics*

Spinneret:

A showerhead-like nozzle used to extrude liquid solution to make fibers.

Bud grafting:

The process of ensuring genetically identical trees to produce a pure product.

Flax:

The plant from which linen is derived, and another name for linen.

Wool classing:

The process of classifying wool into distinct categories.

Desizing:

The process of removing sizing materials from yarns or fabric, usually by application of acid or enzymes.

Bamboo kun:

Bio-agent found in bamboo which gives it anti-bacterial qualities.

Scouring:

The process of removing dirt, animal grease, sand, vegetable matter, and various impurities from certain fibers.

Carbon footprint:

The total set of greenhouse gas (GHG) emissions caused by an organization, event, product, or person.

Xylem:

The woody core of the jute plant.

Exocarp:

The husk of the coconut.

Answer Key

Final Exam Multiple Choice Questions – Part III
Guide to Green Fabrics

1. Which fabric glides like real silk over skin surfaces glowing with a bright luster, which is retained even after continuous washing?

   Modal

2. Bamboo comes from a _______, not a plant.

   grass

3. At one time, negative publicity and a relentless drive against marijuana plants reduced this once thriving plant into a rarely grown crop, especially in the United States. Name the crop.

   Hemp

4. Which two countries lead in the world production of wool?

   Australia and New Zealand

5. Abaca fiber is biodegradable and eco-friendly as abaca is mainly composed of cellulose, _______ and lignin.

   pectin

6. Linen has a wide range of characteristics which makes it preferred wear especially during _________.

   summertime

7. Spider silk is a very strong fiber and it is estimated that a thin rope of spider silk measuring in thickness to a pencil can stop a __________ in motion.

   Boeing 747

8. One of the finest man made fabrics not sourced from synthetic materials bears the name _______, which means *ingredients from the earth.*

   Ingeo

Answer Key | **Extra Credit Assignments # 6 - 19**
Guide to Green Fabrics

 # 6

Q: Describe the viscose (rayon) and mechanical processes of bamboo fiber production and the various stages the plant goes through before it becomes fabric. Be sure to include all stages of production including (but not limited to) pulping, chemical infusion, combing and spinning. Which process is more beneficial for the environment?

A: Bamboo is made from the pulp of the grass and is a cellulose fiber. Producing bamboo requires pulping the leaves and crushing the soft inner pith until the thin fibers separate. Certain patented technologies used for preparing it use a hydrolysis-alkalization process to extract fiber through a chemical process. This is the primary method of bamboo fiber production. Therefore, in most cases it is then soaked in a solution of 15-20% sodium hydroxide, which leads to the formation of alkali cellulose, which is crushed and dried. It is then made into a gel using carbon disulfide, which produces cellulose sodium xanthogenate, to which sodium hydroxide is added to create a rayon-like, viscose solution. This viscose, or bamboo rayon, cellulose is run through showerhead-like spinneret nozzles into containers of diluted sulfuric acid that converts it to bamboo fibers ready to be spun into yarns, which are sometimes bleached. Depending on the size of the yarns bamboo can be extremely thin. However, it has great tenacity and elasticity. It quite resembles cotton at this stage and handles much like cotton yarns. From there, bamboo can be woven or knitted into beautiful, lustrous fabric, which resembles silk depending on the weave. Alternately, there is a mechanical process similar to hemp or flax production which involves crushing the stalks while natural enzymes work to further break down the pulp. The fibers are then combed out and prepared for spinning. The result of the mechanically produced bamboo is a much harder and stiffer fabric, which is not widely marketable for garment production, so it is much less in demand. Both processes were developed in China, the world's leading producer of bamboo. The mechanical process is more beneficial to the environment. Due to its biodegradability, mechanically produced bamboo fabric decomposes into the soil by sunlight and micro-organisms without releasing any harmful gases, such as methane, a harmful by-product of decomposing waste found in dumps and landfills, into the atmosphere.

7

Q: Describe the process of EcoFi® production from sourcing materials to the finished product.

A: The process used to create EcoFi® fabric from plastic bottles begins with sourcing the bottles from local recycling facilities that package and transport it to the processing plant. The sorting process begins with the removal of labels and caps, then the bottles are sorted by color. The bottles are then sterilized before being crushed and chopped into tiny flakes, which find their way into large vats ready to be melted. The process of melting these flakes is carried out with simultaneous stirring until the flakes have been turned into a rich, thick liquid ready to be extruded through a spinneret, which looks much like a shower head. Once extruded through the spinneret, the liquid is transformed into strands of silky polyester. Stretching the fibers allows for strands to be created in various thicknesses, lengths, and strengths. The fibers are then crimped and cut into shorter pieces and readied for shipping. The task of turning these strands into fabric is completed by knitting and weaving.

 # 8

Q: What are the eco-friendly processes used during the production of e-leather?

A: There are several advantageous, eco-conscious process applications used to produce e-leather. These include the use of high-pressure water jets to clean the leather and tangle it, and it is chemical-free. Plus, there is an absence of oil-based adhesives used in conventional bonding, thus reducing our carbon footprint. Clean technology provides the muscle to *loop* the process by regenerating emissions from solvents via thermal oxidation, and this regenerated fuel acts as clean energy in the manufacturing process. Indeed, many stringent regulations are met during these steps. As a bonus, non-toxic water recycling remains a foremost activity used in production, offering a total closed-loop process, which is clean and green.

 # 9

Q: How does fabric made from milk protein aid beauty and personal care?

A: Fabric made from milk protein is not only a reality, but is well on its way to being an answer to many of the skin problems that plague millions of people in a world that is slowly losing its protective ozone layer. Milk protein, as it is known, gives the wearer almost the same benefits that come from ingesting the liquid. These positive benefits are the result of the 18 amino acids contained in milk in its liquid form also being found in the fabric. The amino acids nourish the skin, protecting it from allergens, and reportedly go a long way in postponing the onset of wrinkles.

 # 10

Q: How is cashmere cared for to ensure its luxury appeal remains intact and can be preserved for generations?

A: It is best to keep cashmere away from any obvious conditions that could soil it. In case it gets dirty, washing the soiled area in cold water by hand using a fine, washable soap is best followed by dry cleaning, the preferred cleaning method for cashmere. However, if washing by machine, be sure the machine does not have an agitator during the hand wash or wool cycles. Cashmere should not be wrung-out to get rid of excess water as this may distort the shape of the garment. Instead, it may be rolled in a towel and pressed for excess water, then laid flat to dry, away from heat or direct sunlight. When dry, it should be brushed with the nap. Cashmere can be ironed when it is slightly damp on a cool setting by using a press cloth, which will preserve the fabric's qualities. Ironing should be done on the wrong side, or back side, of the garment. Unfortunately, a hot iron can scorch the fibers permanently. Because cashmere is a protein fiber it should not be washed using any kind of chlorine bleach as this leads to loss of color, permanent yellowing, and weakening of the fiber. In fact, it will dissolve in bleach so it's best to avoid it entirely. The fabric must also be protected against friction if at all possible. For example, it should not be sandwiched between the wearer and any surface, such as an armrest or piece of furniture.

To store cashmere garments, use tissue paper for knitwear lying flat in a drawer to avoid hanging. Woven garments may be hung on padded hangers, and to preserve the shape, buttons and zippers should be fastened. Wrinkles will generally disappear when garments are given a rest. If garments are to be stored for a season, they should be grouped in a loose garment bag and stored in a cool, dry place. Knits may be stored in a drawer and moth crystals or spray will help control moths. As evidenced, extreme care is urged for the preservation of one of the most precious fibers available in the marketplace today.

 # 11

Q: Explain the process of soybean fiber production.

A: The production process begins after tofu and soy milk have been extracted from the soybean. The leftover cake is distilled in order to realize the spherical protein. Next, the protein is converted to liquid through a special process. Under the functioning of an auxiliary agent and biological enzyme, spherical protein's space structure changes and it liquefies. High polymers are added to this mixture and it is cooked. The protein is then passed through a spinneret that looks very much like a showerhead in a process called wet-spinning. The fibers are then stabilized by acetalizing. This liquid is solidified into protein fibers using multiple processes like chemical cross-linking, curving, thermal setting, and cutting.

 # 12

Q: What are the unique qualities of piña that make it extremely expensive and desirable?

A: The leaves produce a silky fiber, which is delicate yet very strong, sometimes even stiff, and white or ivory in color. The fiber is as thin as hair, soft and fine, and is naturally lustrous. Almost see-through, piña is lightweight and has an elegant and beautiful appearance. It has an excellent textural quality that gives it character. It is naturally saltwater resistant and does not easily wear down. Expensive, heirloom-like wedding garments are fashioned from this fiber, which is often blended with silk or polyester. It appears and behaves similar to linen, yet is softer than hemp. Because of its scarcity and hand management techniques, it is considered rare and is very expensive.

 # 13

Q: What are piña's various uses among people of different countries?

A: In the Philippines, the Barong Tagalong, which are formal dresses and wedding dresses, are the most recognized items made from piña. The hand embroidery applied to garments made of piña, called piña calado, place the items at the top of any fashion wish list. The people of the Haiman Island in China have fabricated a coarse variety of the material, which looks similar to grass cloth. Indian shoemakers use piña thread in their manufacturing process while West Africans are known to use it for making caps and capes as well as a stringing material for jewelry. In Guam, fishermen use it to make their nets and also for wrapping cigars. Mats, bags, linens, and table fashions are made of piña. The fibers are also used for making ropes, twine, and paper.

14

Q: Provide extensive and thorough detail regarding products created using recycled nylon.

A: While apparel holds the most important use of recycled nylon with designers using it to design sportswear, the use of recycled nylon in carpet making is also very great. The above-mentioned are just a couple of uses of this multi-faceted fiber whose application ranges from clothing to packaging. Recycled nylon is used extensively in the creation of performance clothing as well as uniforms, working apparel, socks, dress pants, and innerwear such as vests. In the furnishings industry it is used to cater to the demands of the hotel industry as well as hospitals. While the former uses recycled nylon to give itself the tag of *green hotel* by using it for drapes and bed sheets, the latter uses it for making furniture and privacy curtains. It is also used to make sewing thread, zipper tapes, product labels, tote bags, office paneling, and seating, as well as banners and signboards.

15

Q: What is wild silk and what do these silkworms eat?

A: Wild silk, or *Tussah silk* (China) or *Tasar silk* (India), is associated with a more natural way of raising silkworms, such as in natural tree environments with natural movement and eating patterns. These silkworms do not live on mulberry trees exclusively, but roam naturally amongst a variety of trees. Considering that naturally hatched cocoons (where the moth emerges on its own) produce staple fibers in general (because of tearing), it is common practice that wild silk cocoons are plucked from the natural trees. They are then opened to expose the pupa, and it too is normally destroyed by heat or hot water. These cocoons provide reeled silk, or long filament yarns, which are higher in quality and price.

16

Q: Articulate the consumer perception of silk and some of its various end uses.

A: Silk has long been sought after for the production of myriad consumer products of the highest quality. When one hears *silk*, the automatic response is that the associated product has a higher standard and will be of more value simply due to the fiber therein. Silk is used worldwide for men's and women's garments, children's clothes, jackets, scarves, sweaters, hats, and all types of accessories. In household furnishings, it is used to make pillows, luxury throws, and cushion covers, among other luxury items. Printed or painted, silk may even be framed as art.

17

Q: What is the best way to clean garments and other products made of silk?

A: Silk should be washed using a mild soap, and alkaline-based detergents are best avoided as this may weaken the fabric, as does soaking it for a long time or using bleach. After washing and rinsing the garment in cold water, it should be allowed to drip dry or be pressed between towels to remove water.

Silk tends to be weak when wet, so at this stage it must be handled carefully. Adding a capful of vinegar or lemon juice to the final rinse restores the luster that is lost during the washing process. Light ironing from the inside out is recommended. If followed properly, these tips lead to good care and ensure that the garment performs beautifully over time.

 # 18

Q: What accounts for soybean fiber exhibiting properties of both natural and synthetic fibers?

A: Due to a natural material being the main ingredient of the fiber, it retains the elegant luster and comfort of natural fabrics. Because it's produced in association with chemical additives, it takes on some of the qualities apparent in man-made fibers. As such, it has good tensile strength with no shrinkage, even in boiling water, and is easy to care for like most synthetic fabrics. Boasting excellent absorption qualities, soybean protein offers predictable dryness to wearers during hot weather. Soybean protein can be blended with other fibers to showcase a multitude of new fabrics that combine the innate qualities of both.

 # 19

Q: Describe the various processes that convert the flax plant into linen fiber.

A: The flax plant is removed from the ground and deseeded in a process called *rippling*, or even by *winnowing*. Being a bast fiber, the stalks are immersed in water in a process called *retting*. Retting disintegrates the pectin, which is there to hold the fibers together since almost 70% of the stem is cellulosic material. The next step in producing the fiber involves removal of the woody portion of the stem by crushing it between a set of rollers, and releasing the fibers in a process called *scutching*. The fibers are then combed with the *heckling comb*, which removes the short fibers (tow fibers) leaving behind the typically long beautiful flax fibers (line fibers), which are used to create the finest garments and home accessories.

Section 5

Suggested Grading Criteria

Student Final Grade

Grading Criteria | **Suggested Guidelines for Student Final Grade** *Guide to Green Fabrics*

Unit Test/ Final Exam:

Final Exam Questions (Part I) – 70% of final grade
Term Definitions (Part II) – 15% of final grade
Multiple Choice Questions (Part III) – 15% of final grade

Student Workbook Grading Options:

1. Use as percentage of overall grade, per above, such as 5%, and adjust other percentages.

2. Use as extra credit up to a percentage of overall grade, such as 3-5%, which will offset final grade.

Extra Credit Worksheets:

Use as extra credit up to a percentage of overall grade, such as 0.5% per worksheet on a maximum number of worksheets (5-10 worksheets = 2.5-5% of grade), which will offset final grade.